By Dawn Erdos & Leslie Singer

A Subsidiary of
Henry Holt and Co., Inc.

First printing.

ISBN 1-55828-297-1

Printed in the United States of America

10 9 8 7 6 5 4 3 2 1

Senior Development Editor: Cary Sullivan
Book and cover design: Leslie Singer Design, NYC
Cover illustration: Adam Cohen
Indexing: Janey Brand

TO THE COMPUTER AGE
FOR NURTURING OUR BUSINESSES, TEACHING US PATIENCE,
AND CONNECTING US TO THE FUTURE.

GRATEFUL ACKNOWLEDGEMENTS

THE AUTHORS COULD NOT HAVE PRODUCED THIS BOOK WITHOUT THE GENEROUS AND UNFAILING SUPPORT OF MANY INDIVIDUALS. WE WOULD LIKE TO TAKE THIS OPPORTUNITY TO EXTEND OUR MOST SINCERE GRATITUDE TO THEM.

All of the people and companies who generously shared their time and expertise to make this book possible.

Steve Berkowitz, Publisher, for having the vision to publish this book. Cary Sullivan, Senior Development Editor, for her generous support and for going way beyond the call of duty to produce this book. Nettie Aljian, National Accounts Manager, for her valuable input. Fauzia Burke, Marketing Manager, for her enthusiastic support. Erika Putre and Amy Carley for never getting annoyed at us. Diane Cronin for patience, dependability, loyalty, and great design sense. Rosemary Intrieri for putting in her two cents. Joanna Harbert-Doster for her tenacity and dependability. Janey Brand and Siri Gian Khalsa for support and guidance beyond words. Sat Sat Nam Khalsa and Sat Nam Khalsa for their many late nights.

CONTENTS

CONTENTS

CHAPTER 1

CHAPTER 2

CHAPTER 3

CHAPTER 4

Because the computer graphics industry is so diverse in terms of applications, we decided that the best approach for *Concepts in Computer Design: A Professional Perspective* would be to present what other designers and artists are doing and how they're using the latest technology. So many people are producing so many fascinating projects in so many different ways. We felt learning by example would be a much more informative—and interesting—experience for the reader than a typical how-to book about an industry where nothing is typical. Our mission has been to provide practical information, to offer a sampling of the vast array of options currently available to graphic artists and designers, and to instill in the reader a genuine sense of confidence and inspiration.

Since so many people have started—or are thinking of starting—their own businesses, *Concepts in Computer Design* brings you a practical, business perspective of computer technology as it relates to the graphic arts. An ideal resource for people developing start-up operations, this book provides information on how respected industry leaders have structured their businesses around computers. Equipment lists, the benefits of selecting computerized design (such as significant savings in time and costs), the availability of new and exciting

CHAPTER 5

CHAPTER 6

CHAPTER 7

creative tools, budget considerations, client needs, and many other factors guide the reader through the often confusing world of technology with a sense of what can be accomplished.

The rapid-fire pace of technology can often be intimidating, but we believe this book will help you to find your own comfort level, as well as a sense of the advantages available in computer graphics and design. We follow 11 companies and individuals through their creative and production processes. We've selected the widest range of desktop operations in a cross-section of industries. Each has a different view of the computer as a tool, from relying on it as little as possible, to using it as much as possible, and from small single workstation setups, to sprawling multi-million dollar systems.

Chapter 1 follows Royal Viking through the production of a two-color newsletter, produced on a Windows-based PC and a web offset press aboard a luxury cruise ship.

Chapter 2 looks at how *Wired* generates a completely electronic six-color magazine.

Chapter 3 studies how *USA Today* combines traditional techniques with some of the most

CHAPTER 8

CHAPTER 9

CHAPTER 11

CHAPTER 10

sophisticated equipment available to produce a daily, national four-color newspaper.

Chapter 4 covers the steps involved in producing *Concepts in Computer Design: A Professional Perspective.*

Chapter 5 investigates how Diana DeLucia Design executes a corporate identity campaign for Medallion Hotels.

Chapter 6 follows Frankfurt Balkind Partners through the creative process and production of a six-color annual report for Time Warner Inc.

Chapter 7 illustrates how Rucker Huggins Design developed the packaging for Fractal Design Painter.

Chapter 8 tags along with the folks at Ben & Jerry's Ice Cream through the production of a brochure.

Chapter 9 discusses how The Color Space produces large-format digital output.

Chapter 10 covers how Media Designs created ad slicks for an advertising campaign for the Associated Press.

Chapter 11 looks at how Adam Cohen uses Photoshop to illustrate a series of murals for a chain of record stores.

CHAPTER 1

NEWSLETTER

This chapter discusses the production of a daily newsletter for Royal Viking Line passengers. Learn how one woman uses a personal computer to produce this three-color publication while at sea.

Royal Viking *Skald*

ROYAL VIKING LINE

THE WORLD'S FINEST

ROYAL VIKING

SKALD

ROYAL VIKING SUN • VOL. 77 • NO. 16 • MONDAY • DECEMBER 14, 1992
CRUISE: EN ROUTE TO GREAT STIRRUP CAY, BERRY ISLAND, BAHAMAS

This painting, by passenger Anne Vose, will be one of the many on display today in the Midnight Sun Lounge foyer.

PASSENGER ART EXHIBIT TODAY

More than 200 paintings will be on display today in the Midnight Sun Lounge foyer from 9 a.m. to 9 p.m. today. They are the work of fellow passengers who have been attending Eljay Torger's art class every day at sea. The show features works by both novice and experienced artists.

Media used include pencil, watercolors, pastels, pen and ink and tissue paper collage. The ancient art of Oriental brush painting will also be represented. In addition, works for the Needlepoint Club will be displayed.

Eljay will suspend her normal day of instruction to give a lecture on the works displayed, beginning at 3 p.m. in the Midnight Sun Lounge foyer. All passengers are invited to attend.

WORLD AFFAIRS CERTIFICATE PRESENTED BY DR. HARM DEBLIJ

A GRAND FAREWELL BALL FOR THE CIRCLE SOUTH AMERICA CRUISE

AN EVENING WITH MICKEY ROONEY TONIGHT IN THE NORWAY LOUNGE

The Royal Viking Skald is a daily three-color newsletter produced entirely at sea for the passengers of the Royal Viking Sun.

INTRODUCTION

Royal Viking Skald is a six- to eight-page daily newsletter produced for passengers aboard the Royal Viking Sun luxury cruise ship. One of the most fascinating facts about this newsletter is that it's a three-color piece, produced and printed on a ship while in the middle of the Mediterranean Sea.

The Skald is put out by one person, Angela Linsey, who is responsible for developing and writing the features and stories, creating graphics elements, retouching photographs, inputting and formatting text, and designing each issue. She is assisted by a systems manager who takes care of any technical issues that arise.

Royal Viking Line has had its electronic publishing system in place for nearly two years. Prior to that, the newsletter was produced on board using traditional paste-up methods. Says Angela:

> *Basically, they were doing it manually. They had some sort of a computer that would typeset but then if there were any errors they'd have to print out the corrections and paste it on. Like they did newspapers.*

ROYAL VIKING

SKALD

ROYAL VIKING SUN • VOL. 81 • NO. 5 • SATURDAY • MARCH 20, 1993
DOCK/SAIL: ISTANBUL, TURKEY

Istanbul's Mosques

Istanbul is the home of approximately 500 mosques, some of which are among the world's most beautiful. With towering minarets, majestic domes and solemn worshippers, the religious structures add much to the city's exotic flavor and scenic skyline.

A mosque is a place of public worship for members of the Moslem faith. Comprised of a rectangular courtyard, which usually contains porticos and a fountain, they house halls for meetings or common prayer. One or more minarets – slender towers from which worshippers are called to prayer – rise outside the central building. A mosque may have as many as six minarets, but only those built by sultans are permitted to have more than one.

The interior of a mosque is richly carpeted but unfurnished, with no altars or statues. An arched niche called a "mihrab" faces the direction of the holy city of Mecca. Turkish mosques are characterized by beautiful tile wall decorations depicting abstract forms.

Five times a day, religious officials called "Muezzins" climb the steps of minarets to call the faithful to prayer. At the mosque fountain, worshippers wash their faces, hands and feet before going inside to pray. Shoes are never worn inside a mosque.

While worshiping, the faithful prostrate themselves, touch their foreheads to the floor. Moslem services are conducted in the ancient Arabic tongue.

Most of Istanbul's mosques are located around the Golden Horn. Anyone may go into a mosque, even during prayer sessions. Visitors must be very quiet, however. Photography is permitted, but for courtesy forbids the use of flash cameras. Shoes, of course, are removed before entering the mosque.

SHOWTIME WITH THE HUBER MARIONETTES,
BOUZOUKI PLAYER CHRIS SARANTAKIS AND DANCE TEAM ALAN & DAWN

LATE-NIGHT GAME SHOW — "NAME THAT TUNE"

The front page of each issue frequently features information about the current port-of-call.

WHAT'S IN AN ISSUE?

Passengers on the Royal Viking Sun all have the same response to the newsletter: They love it. One passenger, for instance, liked waking up at 8 AM, as the ship was docking at Gibraltar, to find the newsletter with information about that port. She would not only be prepared for her onshore excursion, but she would be even more eager and interested in the day's events. Passengers also are impressed with the way newsletter articles are coordinated with all of the ship's daily events, as well as with the Royal Viking Sun's morning television show, *Sunrise.*

The front page of each issue is usually dedicated to a feature about one of the cruise destinations, highlighted cruise event, or celebrities who are on board the ship. Previous articles have included features about France's Cote d'Azur, Venetian heritage, Istanbul's mosques, and shipboard celebrities such as journalist Charles Kuralt and Broadway star Carol Lawrence.

Page 2 of the newsletter contains shore excursion and chief purser information. The center pages are reserved for daily program information,

ON BOARD THE ROYAL VIKING SUN, THE WINE IS AS IMPORTANT AS THE FOOD

As every bon vivant knows, fine food deserves fine wine – and that includes the five-star-plus cuisine served on board the *Royal Viking Sun*.

The Royal Viking Dining Room's wine cellar boasts a selection of some 130 wines, from various regions of France, Italy, Germany, Spain, the United States and Australia.

Selecting these wines shoreside is Royal Viking Line's own wine consultant who travels the world picking out the best labels and vintages for our discriminating passengers. Recently, Cornell Professor Stephen A. Mutkoski joined our panel of experts to compile a list of wines from our cellar that he felt would best complement and enhance each night's meal.

On the *Royal Viking Sun*, Food and Beverage Manager Johannes Lindthaler and Provisions Master Horst Katschnig coordinate ordering for the cellar and strive to see that a plentiful supply of nearly 14,000 bottles is kept stocked in the wine store on "B" Deck.

Head Sommelier Brian O'Brien and his wine stewards are at your service when it comes to making a selection to perfectly complement your meal. It is also their duty to serve wine at the proper temperature for optimum enjoyment.

To ensure the best service possible, Brian and the wine stewards recommend that you select your wine for dinner at lunch, so that your bottle will be at the proper temperature at mealtime. (Due to the high demand of some wines, we apologize if the vintage or shipper varies from our wine list.) Brian notes that the most popular wines on board are California whites, as well as French chablis. However, he suggests that passengers might like to try a red wine. As he puts it, "Most people seem to prefer white wines, but many connoisseurs seem to prefer reds."

If you'd like to taste a new wine this cruise (and what better opportunity to experiment?) Brian recommends a few of his personal favorites that he thinks you'll enjoy and has provided some notes on their characteristics:

Louis Roederer, Cristal (1980)

Perhaps the most stylish and elegant of all champagnes, the '83 vintage leaves no doubt as to its pedigree – dry, deeply flavored and a perfect, crisp finish make the Cristal a graceful addition to any dinner table.

White Wines

Chateau Montelena (1989)

Since 1969, Chateau Montelena has had an enviable track record of producing some of the best chardonnays in

California, if not the world. The '89 offering does not disappoint; its oaky, buttery flavor and full body make it an ideal wine to combine with a richly flavored fish course.

Corton-Charlemagne, Grand Cru (1987)

One of the legendary great white burgundies and very keenly priced, the '87 is an atypical Corton-Charlemagne – vigorous and powerful with a distinctive nutty vanilla-oak flavor. A wine to savor, to linger over and to remember long after the last glass has been poured.

Red Wines

Chateau Leoville-Lascases (1983)

Probably the most highly regarded of St. Julien wines – with ample justification. Its deep color and intense bouquet harmonize with a lingering finish to make a wine that is all grace, elegance and distinction.

Opus One (1988)

A Mondavi/de Rothschild joint venture – where the old world comes together with the new to produce a Pauillac-inspired blend of Cabernet Sauvignon, Cabernet Franc and Merlot that epitomizes the best of contemporary Californian winemaking. Its intense fruit and bouquet that evolve almost magically in the glass make Opus One a Californian masterpiece.

PROFILE

DANCE TEAM ALAN & DAWN

Dance partners since their early teens, our dance team of Alan and Dawn are marriage partners as well. Both hail from Great Britain – Alan's a "country lad" from Worcester; Dawn's a "city girl" from Birmingham. Alan and Dawn were veterans of numerous amateur dance championships before turning professional in 1979. Their amateur credits include winning many British Championships. They represented Britain in world and international competitions, and reigned for three years as champions of the British TV show "Come Dancing."

They have taught, lectured and performed throughout Europe, operated a successful dance studio in England and have been cruising for seven years.

As competitors, they had specialized in Latin American dances but, upon advice from their coach, have since introduced a diverse range of styles into their cabaret-style act. Their repertoire now includes everything from the Mambo to the Waltz, from the Charleston to the Hoe Down. Alan's favorite is the lively, swinging Jitterbug; Dawn prefers to dance the romantic Rumba. Alan and Dawn admit they're happiest when they're dancing – whether they're teaching dance or performing. "Dancing is so much fun," Dawn says. "It keeps the mind active, the muscles toned and it's great socially."

In addition to performing their intricate routines in Norway Lounge variety shows, Alan and Dawn also offer a dance class each day at sea. Check your Daily Program for time and location.

IN THE LIMELIGHT

THE HUBER MARIONETTES

One of America's foremost artists in the realm of puppetry, Phillip Huber is world-renowned for his "Art in Entertainment." This Emmy-award winning performer has been perfecting his craft since the age of 3. Primarily self-taught, Mr. Huber designs, costumes and builds all his characters placing emphasis on detail to create a total illusion.

In 1980, Mr. Huber and David Alexander (director and business manager) created the Huber Marionettes, a company dedicated to pursuing the art of puppetry while entertaining the audience. Their variety format incorporates many aspects of theater, transcending all age and language barriers.

The Huber Marionettes have performed before heads of state during national tours of Japan, have opened for celebrities including Donald O'Conner and Carol Lawrence, have appeared on numerous television commercials and specials, for example, Walt Disney's "New Vaudevillians Two;" "John Denver and the Muppets: A Christmas Together;" and, most recently, NBC's "The Tonight Show" with Jay Leno. Southern Californians might recognize the Huber Marionettes from their night club appearances at Pasadena's "Ice House" and Newport Beach's "Magic Island." In addition, they have the honor of being the first – and only – marionette act to perform and lecture at Hollywood's famed Magic Castle.

OUR BAHAMIAN STAMPS ARE EASY AND CONVENIENT

The last thing anyone wants to do on a once-in-a-lifetime trip to an exotic country is wait in line at a not-so-exotic foreign post office -- use our Bahamian stamps.

Thanks to an international agreement with the Postal Union of Lausanne, Switzerland, passengers on the *Royal Viking Sun* may use Bahamian stamps when the ship is in international waters (the *Sun* carries Bahamian registry). These stamps are cancelled on board with our own postmark and landed at the next port of call.

Please bring cards and letters to the Reception Desk on Norway Deck (#8) by 6 p.m. the day prior to our arrival at the port from which you wish the mail posted.

Bahamian stamps are for sale (cash only, please) at the Reception Desk and prices (in U.S. currency) are as follows:

To North America
Postcard $0.40
Letter 0.50
To Europe & South America
Postcard $0.45
Letter 0.50
To Africa, Asia & Australia
Postcard $0.50
Letter 0.60

Please note: *Due to international law, we are not allowed to sell stamps while the ship is at port.*

One of the newsletter's interior sections includes shore excursion, purser, daily program, and special events information.

and the back page features daily ship-board services and hours. The remainder of the newsletter contains features and profiles about cruise destinations, guests, and ship staff members.

Issues are produced daily, but an issue doesn't simply start at the beginning of a day. Angela actually starts an issue two days before it's due to come out. Basically, each newsletter follows the same plan:

1. Research and writing.
2. Photography.
3. Page composition.
4. Output and printing.

ELEMENTS OF THE PROCESS

Let's take a look at these steps in a little greater detail.

Research and Writing

Information is compiled beginning two days ahead of time. The stories are written, typed into Microsoft Word for Windows, and stories are checked using Grammatik.

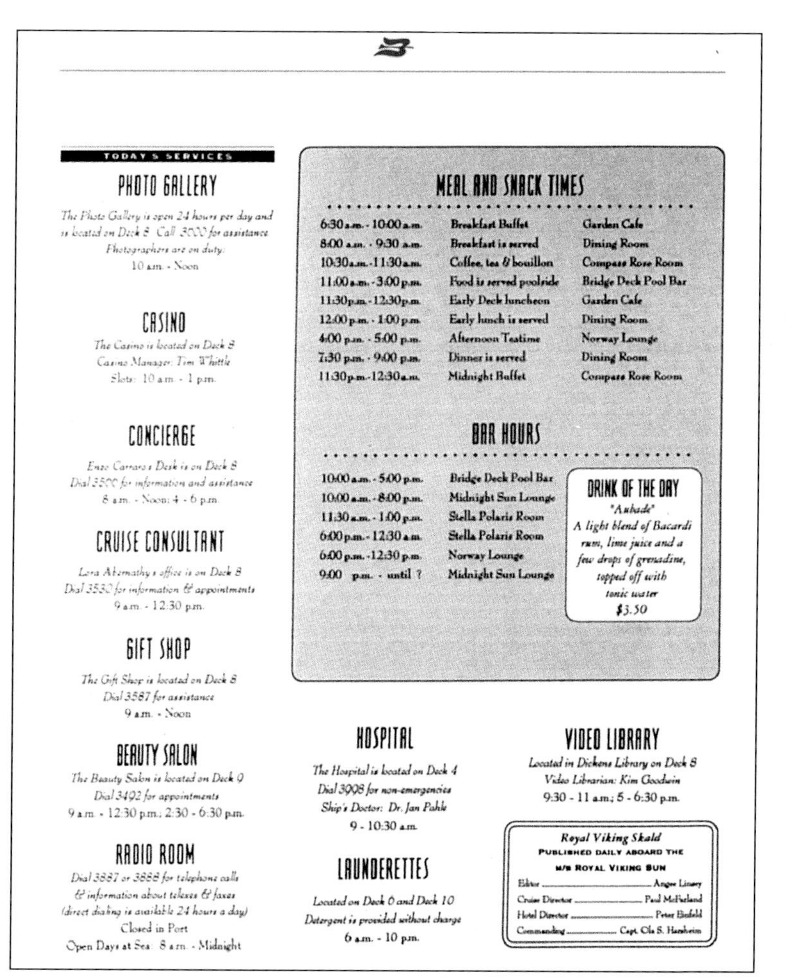

TODAY'S SERVICES

PHOTO GALLERY

The Photo Gallery is open 24 hours per day and is located on Deck 8. Call 3000 for assistance.
Photographers are on duty:
10 a.m. - Noon

CASINO

The Casino is located on Deck 8
Casino Manager: Tim Whittle
Slots: 10 a.m. - 1 p.m.

CONCIERGE

Enzo Carraro's Desk is on Deck 8
Dial 3500 for information and assistance
8 a.m. - Noon; 4 - 6 p.m.

CRUISE CONSULTANT

Lora Abernathy's office is on Deck 8
Dial 3530 for information & appointments
9 a.m. - 12:30 p.m.

GIFT SHOP

The Gift Shop is located on Deck 8
Dial 3587 for assistance
9 a.m. - Noon

BEAUTY SALON

The Beauty Salon is located on Deck 9
Dial 3492 for appointments
9 a.m. - 12:30 p.m.; 2:30 - 6:30 p.m.

RADIO ROOM

Dial 3887 or 3888 for telephone calls & information about telexes & faxes (direct dialing is available 24 hours a day)
Closed in Port
Open Days at Sea: 8 a.m. - Midnight

MEAL AND SNACK TIMES

6:30 a.m. - 10:00 a.m.	Breakfast Buffet	Garden Cafe
8:00 a.m. - 9:30 a.m.	Breakfast is served	Dining Room
10:30 a.m. - 11:30 a.m.	Coffee, tea & bouillon	Compass Rose Room
11:00 a.m. - 3:00 p.m.	Food is served poolside	Bridge Deck Pool Bar
11:30 p.m. - 12:30 p.m.	Early Deck luncheon	Garden Cafe
12:00 p.m. - 1:00 p.m.	Early lunch is served	Dining Room
4:00 p.m. - 5:00 p.m.	Afternoon Teatime	Norway Lounge
7:30 p.m. - 9:00 p.m.	Dinner is served	Dining Room
11:30 p.m. - 12:30 a.m.	Midnight Buffet	Compass Rose Room

BAR HOURS

10:00 a.m. - 5:00 p.m.	Bridge Deck Pool Bar
10:00 a.m. - 8:00 p.m.	Midnight Sun Lounge
11:30 a.m. - 1:00 p.m.	Stella Polaris Room
6:00 p.m. - 12:30 a.m.	Stella Polaris Room
6:00 p.m. - 12:30 p.m.	Norway Lounge
9:00 p.m. - until ?	Midnight Sun Lounge

DRINK OF THE DAY

"Aubade"
A light blend of Bacardi rum, lime juice and a few drops of grenadine, topped off with tonic water
$3.50

HOSPITAL

The Hospital is located on Deck 4
Dial 3998 for non-emergencies
Ship's Doctor: Dr. Jan Pahle
9 - 10:30 a.m.

LAUNDERETTES

Located on Deck 6 and Deck 10
Detergent is provided without charge
6 a.m. - 10 p.m.

VIDEO LIBRARY

Located in Dickens Library on Deck 8
Video Librarian: Kim Goodwin
9:30 - 11 a.m.; 5 - 6:30 p.m.

Royal Viking Skald
PUBLISHED DAILY ABOARD THE
M/S ROYAL VIKING SUN

Editor .... Angee Linsey
Cruise Director .... Paul McFarland
Hotel Director .... Peter [illegible]
Commanding .... Capt. Ola S. Harsheim

Skald editor, Angela Linsey, begins gathering information two days before an issue is distributed. She then lays out the pages, and sends the proofs to the appropriate departments for review and last-minute additions.

First, Angela gathers and inputs all arrival and departure information if the ship is going to be in port. The shore excursion manager provides her with port activities, such as a port talk, a lecture in the lounge, or a closing of tour sales for a certain port. The Cruise Director fills Angela in on the on-board activities and services. She describes the process:

> *I just keep the pages in the system and then I alter them. And then I also do a first draft of the daily program—for in-port it's usually one to two pages, and if we're at sea it's usually three pages of activities.*

Activities are defined right down to the hour, so Angela composes these pages one or two days ahead of time, then prints them out the night before the print date. She then distributes them to their respective departments for review and last-minute additions. Angela usually interviews any

The Royal Viking has a ship-board photographer who provides Angela with plenty of visual material.

new entertainers on board a day or two before the issue is due out. The morning before an issue is distributed, Angela either finishes the cover story or, if the ship has been to the same port recently, she reworks an existing story from her archives.

Photography

The Royal Viking Sun has a ship photographer who takes photos of ports, guests, celebrities, and other items relating to newsletter features. A lot of the biography photos Angela uses are brought by entertainers who have had them professionally taken on shore. She also has some stock photos available to her:

We have so many entertainers and staff members that are on board

ROYAL VIKING
SKALD

ROYAL VIKING SUN

WELCOME ABOARD!

Captain Ola S. Harsheim (pictured, above), the officers, staff and crew of the *Royal Viking Sun* are happy to welcome you aboard. The ship will be your home in the days ahead, and the pace will be of your own choosing – fun-loving with parties, games and shows, or relaxed, with plenty of time to enjoy the changing scenery. We look forward to serving you, and hope you will enjoy your cruise with us. This special "Welcome Aboard" issue of the *Skald*, our daily newsletter, will introduce you to some of the features and services the *Royal Viking Sun* has to offer.

Angela uses between three and six photos per issue. The pictures are scanned in using a Hewlett Packard Scan-Jet IIc.

frequently enough that we want to keep those photos in our file system. We could put them on disks but some of those photos take up a lot of space. You could put two photos on a disk. So I have a lot of them on the hard drive. Especially the ones that I use all the time. That takes up more than half of my hard drive memory.

Once she has all of the available photos in front of her, Angela selects three to six of them. The photos are then scanned into 200 dpi TIFF files using a Hewlett Packard Scan-Jet IIc with Deskscan software, and edited using

ROYAL VIKING

SKALD

ROYAL VIKING SUN • VOL. 77 • NO. 7 • SATURDAY • DECEMBER 4, 1992
CRUISE: EN ROUTE TO RECIFE, BRAZIL, SOUTH AMERICA

CAROL LAWRENCE

Television, film and Broadway star Carol Lawrence will take the stage in the Norway Lounge at 10 p.m. tonight.

Carol Lawrence has been lighting up the stage and charming audiences since she was 6 years old when she found her knack for performing during the Thanksgiving play at her elementary school in hometown Melrose Park, Illinois. That is about the time she began to study song and dance seriously – the beginning of her quest for Broadway stardom.

At the tender age of 12, Ms. Lawrence won a four-year scholarship at a Chicago dance studio where owner and mentor Edna MacRae took Carol under her wing and began to grooming her for the big stage. Within only one year, she was singing and dancing professionally during the summers.

Ms. Lawrence finished school and quickly found her way to New York City to make her glorious dreams of Broadway a reality. At 20 years old, she landed the key role of her career – Maria in *West Side Story*. The show became a smash hit and she hasn't stopped since.

Her stardom came with the golden age of television where she began to do guest appearances with Ed Sullivan, Gary Moore, Perry Como and the Hallmark Playhouse. "It was an wonderful learning experience for me," Carol says. "I was so young and hadn't really studied that much. I was learning on the job with

Continued inside

WORLD AFFAIRS LECTURES WITH CHARLES SARGENT AND THOMAS DODD

EARLY EVENING CLASSICAL CONCERT SCHOTTEN AND COLLIER

LATE-NIGHT FUN WITH THE LIAR'S CLUB IN THE MIDNIGHT SUN LOUNGE

Ångela frequently features big-name entertainers or passengers for the front page of the newsletter.

Publisher's Paintbrush. Often, the photos need some adjustments—such as improving brightness, contrast, or intensity—to make them look professional in print, which are done in Publisher's Paintbrush.

Page Composition

Text and graphics are imported into Aldus PageMaker 4.0 for Windows, where the text is formatted and wrapped around the graphic elements.

The pages are generated in Aldus PageMaker using several different files, to simplify information management. Angela finds it easier to update old files, rather than work with a PageMaker template. For instance, says Angela:

> *The meal hour page is on a separate document and I have port meal hours, sea day meal hours, port morning meal hours, port afternoon meal hours because they generally are very similar. So I have*

Daily Program

	Good Afternoon	
Midday	Noontime Update with Cruise Director Doug Jones	Public Address
12:00	Midday Melodies with John Mentis at the piano	Midnight Sun Lounge
12:00	Dance Class with Alan & Dawn	Norway Lounge
12:15	Daily Log Update – Join First Officer Nels Forberg	Compass Rose Room
12:15	Karen Reveals today's riddle answers	Midnight Sun Lounge
2:00 - 4:00	Golf Simulator Play – sign up with Golf Pro Ben	Pebble Beach
2:00	Duplicate & Party Bridge with Carol & Norman McCaskill	Card Room
2:00	The Gaming Tables are open for play	Casino
2:00	Free Gaming Instruction in the Casino	Casino
2:15	Afternoon Movie: *Beethoven*; starring Charles Grodin and Bonnie Hunt; Comedy; Rated PG; 1 hr., 28 mins.	Starlight Theatre
2:30	Ladies Quoits Tournament; meet Laurie	Norway Deck, aft
3:00	Men's Doubles Ping Pong Tournament with Chris	Bridge Deck
3:00	Aquacize with Fitness Director Loren	Lap Pool
3:00	Art Class with Eljay Torger – Painting Tropical Scenes	Garden Cafe
3:00	Travel Photography with Mark Torger – Photographing Recife	Dining Room, forward
3:15	Beginners Napkin Folding with Hostess Karen	Dining Room, aft
3:45	Bingo Cards are available; one card for $7; three cards for $15	Norway Lounge
3:45	Teatime Melodies with John at the piano	Norway Lounge
4:00	Galley Tour – Please meet at the Maitre D's desk	Dining Room, aft
4:00 - 5:00	Golf lessons – sign up with Golf Pro Bruce	Pebble Beach
4:00	Harpist Juli sets the mood for Elegant Teatime	Stella Polaris Room
4:00 - 5:00	Afternoon Teatime; finger sandwiches & sweets are served	Norway Lounge
4:00	Hostess Karen reveals the answers to the Daily Quiz	Norway Lounge
4:10	Bingo Bonanza – Join the Cruise Staff & Our Guest Hosts	Norway Lounge
4:30 - 5:30	Golf Clinic "Putting" with Golf Pro Ben	Golf Net
4:45	High/Low Impact Aerobics with Fitness Director Loren	Health Spa
5:00 - 6:30	Video Library is open with Marj	Dickens Library
5:45	"The 5:45 Live Show" – Join Cruise Director Doug Jones and guest Bobbi Baker for comedy, music & the Prize Man.	Cabin TV Ch. 5

Daily Program

	Good Evening	
6:15	Jewish Sabbath Eve Service with Rabbi Arthur Lelyveld	Starlight Theatre
6:15 - 6:50	Dancing to the Jack DeLong Orchestra	Norway Lounge
6:45	Dancing to the Sun Quartet	Midnight Sun Lounge
6:45	Harpist Juli plays for your cocktail hour enjoyment	Stella Polaris Room
7:00	**Classical Concert with Yizhak Schotten (Viola) and Katherine Collier (Piano); featuring the works of Haydn, Beethoven, Granados, Liszt and Schumann**	Norway Lounge
9:00	Luck be a lady tonight! The Casino is open for gaming pleasure!	Casino
9:15	Dance Date with our Hosts to the Jack DeLong Orchestra	Norway Lounge
9:45	Evening Movie: *Lethal Weapon 3*; starring Mel Gibson and Danny Glover Action; Rated R; 2 hr., 2 mins.	Starlight Theatre
10:00	**A Star in the Musical Spotlight – An Evening Carol Lawrence Musical Director Ron Harris conducting the Jack DeLong Orchestra**	Norway Lounge
10:30	Dancing to The Sun Quartet with our Guest Hosts	Midnight Sun Lounge
11:00	Music and Dancing to the Jack DeLong Orchestra	Norway Lounge
11:00	**Game Show fun with The Liars Club with Guest Fibbers: Comedienne Bobbi Baker; Singer David Reign; Magician Fred Becker and Cruise Director Doug Jones Hosted by Chip Hoehler and Karen Kuttner; Scorekeeper Loren Serdar**	Midnight Sun Lounge
11:00	Enjoy a nightcap as harpist Juli plays	Stella Polaris Room
11:30	Midnight Buffet is served as John plays piano	Compass Rose Room
11:30	Dance the Night Away to the Sun Quartet	Midnight Sun Lounge

Late-Night Smile:
The amount of sleep required by a person is usually about . . . 10 minutes more!
Sleep Well –
Cruise Director Doug Jones

Passengers of the newsletter rely heavily on it to inform them of daily events.

four pages and I call that my meal hours page, and then I just fix that. What I do is I copy an old issue and I just erase nearly everything. I just start over because it's all formatted. It's not the most logical way to do it but it works. It's just simple enough to do anew every time.

Once the text and graphics are merged in PageMaker, Angela proofs the newsletter. Once finalized, she delivers it to the ship's printer.

Output and Printing

One of the most striking features of the *Skald* is its three-color format. The printing process for the newsletter is actually quite simple.

The final copy is output using an IBM Laserprinter 10. Angela submits her final laser prints to the printer by 2:00 PM the day before it is to be distributed. The printer shoots film from the laser copy using an Agfa Gavaert Repromaster 1810 with

ROYAL VIKING

SKALD

Although the ship-board printing press has four-color printing capabilities, the masthead and standard color elements are printed on shore. The blue and red masthead is then passed through the on-board press once, using black ink.

either a 185 mm lens or a 135 mm lens. Once the film is developed, plates are etched and placed on an AB Dick 9810XC offset press, which can handle sheets up to 13" x 19.2" x .005".

Although the ship-board offest press has four-color capability, Royal Viking Line has a two-color masthead printed at a shop on shore. This blue and red masthead is then passed through the on-board press once, usually using black ink. This saves Royal Viking both time and money, while presenting a sophisticated image to its passengers.

The Royal Viking Sun is a 673-foot ship with several dining rooms, a casino, spa, theatre, and many other amenities.

THE ROYAL VIKING SUN

The Royal Viking Sun is a mid-size vessel with large-scale amenities. The 673-foot ship has features that include several dining rooms, a casino, a spa, a theater, a card room, two swimming pools, a beauty salon, a photo gallery, and extensive sports facilities. Royal Viking has a unique "guest chef" program that provides its restaurants with menus by some of the most famous culinary personalities.

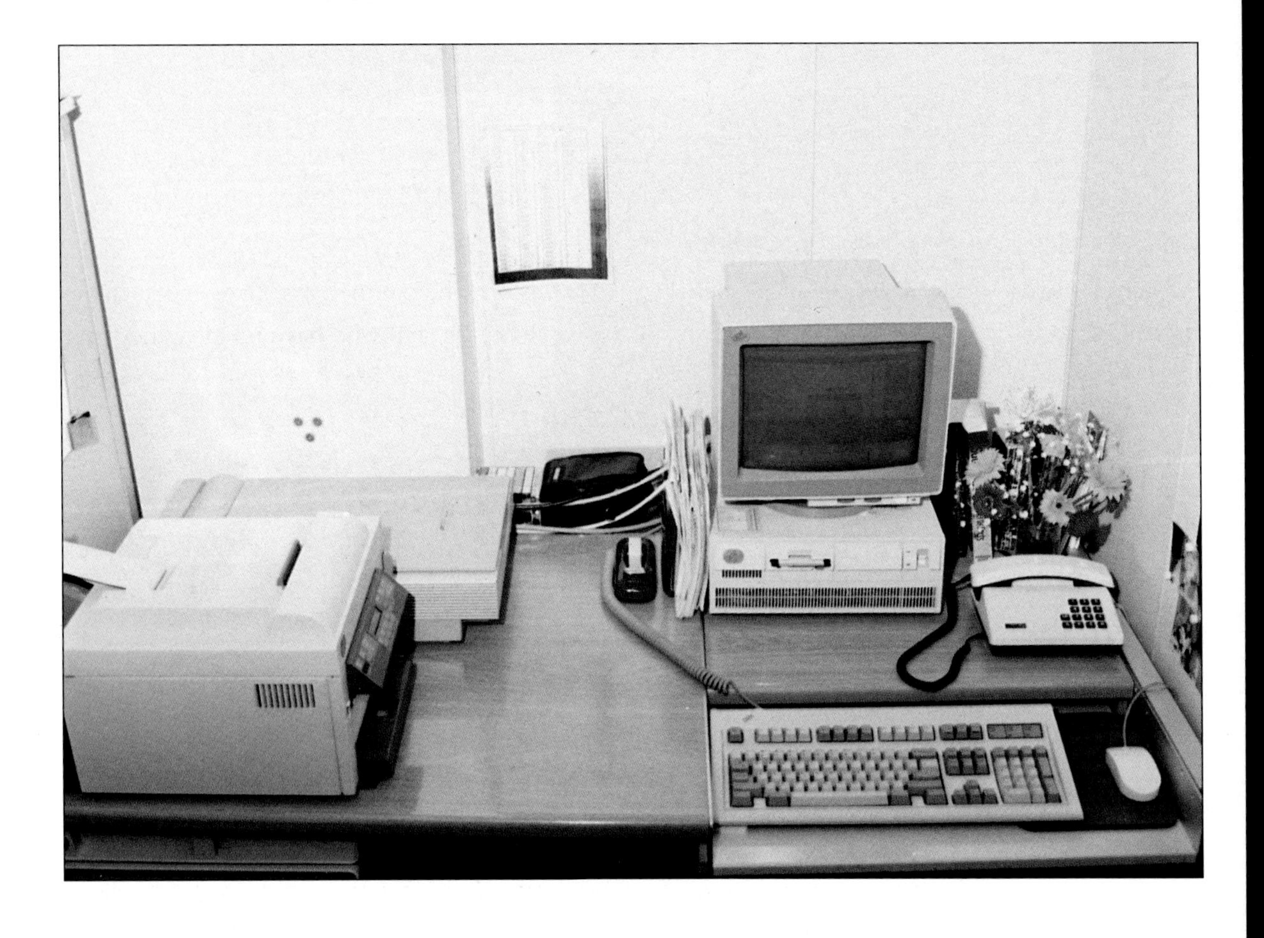

EQUIPMENT

- IBM PS/2 Model 70 386
- Hewlett Packard Scan-Jet IIc
- IBM Laserprinter 10
- Agfa Gavaert Repromaster 1810 camera with 185 mm and 135 mm lenses
- AB Dick 9810XC offset printing press with a plate size of 13" x 19.2" x .005"
- Microsoft Windows 3.1
- Microsoft Word for Windows
- Aldus PageMaker 4.0
- Deskscan scanning software
- Publisher's Paintbrush image editing software
- Grammatik grammar-checking software

CONVERSATIONS

ANGELA LINSEY
Skald Editor

Angela produces a six- to eight-page daily newsletter for the Royal Viking Sun passengers. Prior to working for Royal Viking Line, Angela worked in public relations at Fleishman-Hillard and Dow Chemical. A graduate of the University of Missouri School of Journalism, Angela is also a former Army journalist and National Guardsperson, and an active Naval Reserve Public Affairs Officer.

I do a six- to eight-page daily newsletter for the ship's passengers. The basic purpose of the newsletter is to inform passengers of shipboard information, customs information, port information, and daily activities. We also do bios on the entertainers that come on board and key staff members, along with shipboard information like the hours that services are open and little blurbs about the services on board.

My friends with journalism backgrounds are very impressed with the newsletter because they know how much work it is to just do six to eight pages a day, by yourself, with no English-speaking people reading your material—you're the only set of eyes on it before the passengers see it. But to the people on the ship who have no idea, they think that it's perfectly acceptable and they don't think that it's a big deal.

I find that passengers—Royal Viking passengers are mostly repeat passengers—have grown accustomed to this newsletter. They expect certain things from it. As a new person, I was trained by someone who didn't tell me all those things. So I learned a lot by virtue of the fact that I left things out that passengers expected. They read it every single day. Cover to cover.

They expect the port features, and they like to see the bios on people. They like to have little language blurbs. When we're in Italy, they like some Italian so that when they get into a taxicab they can get to where they want to be. And I didn't do that at first, so I had to start finding people who could help me with the languages.

They really do look at it, and they look for information that will help them on their trip.

RON HOOK
Executive Chef, Doral Saturnia International Spa Resort, Miami, and passenger on the Royal Viking Sun.

Ron has been a chef for 18 years. For the first 12 years he was a traditional chef. Since he started working at the spa six years ago, he has been cooking and eating only spa food. The Doral Saturnia International Spa Resort is a 48-suite destination resort spa, says Ron, "Sort of a cruise ship on land."

I was invited as guest chef for 1992. The line invites selected chefs from around the world and features a menu that is indicative of that style of cooking—mine's spa cooking. They then feature that menu along with the regular menu for the day.

I also was expected to do cooking demonstrations, be interviewed for the on-board TV show and the newsletter, and do a small dinner party.

The newsletter interview was really, really good. In fact, it came out so well that I use it as part of my press kit. It was very well written, and it gave a different feel—it wasn't just a strict bio. It had a picture, a good typeface. It was pretty surprising to see that in a daily newsletter. They could have gone for something simpler, but it really was good copy and well laid out.

It didn't have faded corners or broken letters—it was much nicer than I expected, and much better than I thought was necessary, but it was such a nice surprise.

SUMMARY

Passengers on the Royal Viking Sun can go to sleep on the high seas, and literally wake up as the ship is floating into port—and have at their doors information about the port at which they are docking. Daily delivery of this newsletter is neither a simple undertaking, nor an inexpensive one. But it is one of those extra courtesies that makes full service in the hospitality industry even more hospitable. Unlike a magazine, advertisement, or even a subscription-based newsletter, this is not an investment on which Royal Viking Line can see a tangible return. But the benefit is that the passengers rely on it, are pleasantly surprised by it, and definitely remember it. They also know that Royal Viking Line goes the extra mile for them. And that makes Royal Viking Line stand out in the industry—that's the return on their investment.

CHAPTER 2

MAGAZINE

This chapter gives you an overview of the production of a six-color digital magazine. Learn how *Wired* not only uses desktop technology, but how it reflects the impact of technology on our lives.

WIRED MAGAZINE

WIRED
Jaron Lanier Moves On
3DO - Hip or Hype?
Seymour Papert:
Literacy Is Obsolete
May/June 1993
Rebels with a Cause
(Your Privacy)
$4.95 / Canada 5.95

your education,

your family,

your neighborhood,

As the first mass market product focusing on the effect of technology on our culture, the premier issue of Wired *presented a great leap for the magazine publishing industry.*

INTRODUCTION

Wired magazine is designed to present the issues surrounding technological issues to a general audience, yet it's not merely a trade publication. Since its premier in early 1993, it's focus has been on the larger implication of technology on our society, our lives, and our future, which it does with no holds barred. The subject matter is quite striking, as is the design. With powerful—sometimes disconcerting—images, staggering colors, and rule-breaking design, this magazine is not for the faint-hearted. If you're looking for stories that simply review new digital products, move on. But if you are interested in what the technology means, what it does, and where it is going to lead us, keep reading.

Starting Up

The founders of *Wired* started looking for investment money the same week that the United States attacked Iraq—not the most stable economic environment. It took a while for them to find people who understood the idea and believed in it enough to invest their own money.

This all came together when they spoke to Nicholas Negroponte, creator of the MIT Media Lab. He immediately understood the concept, thought it

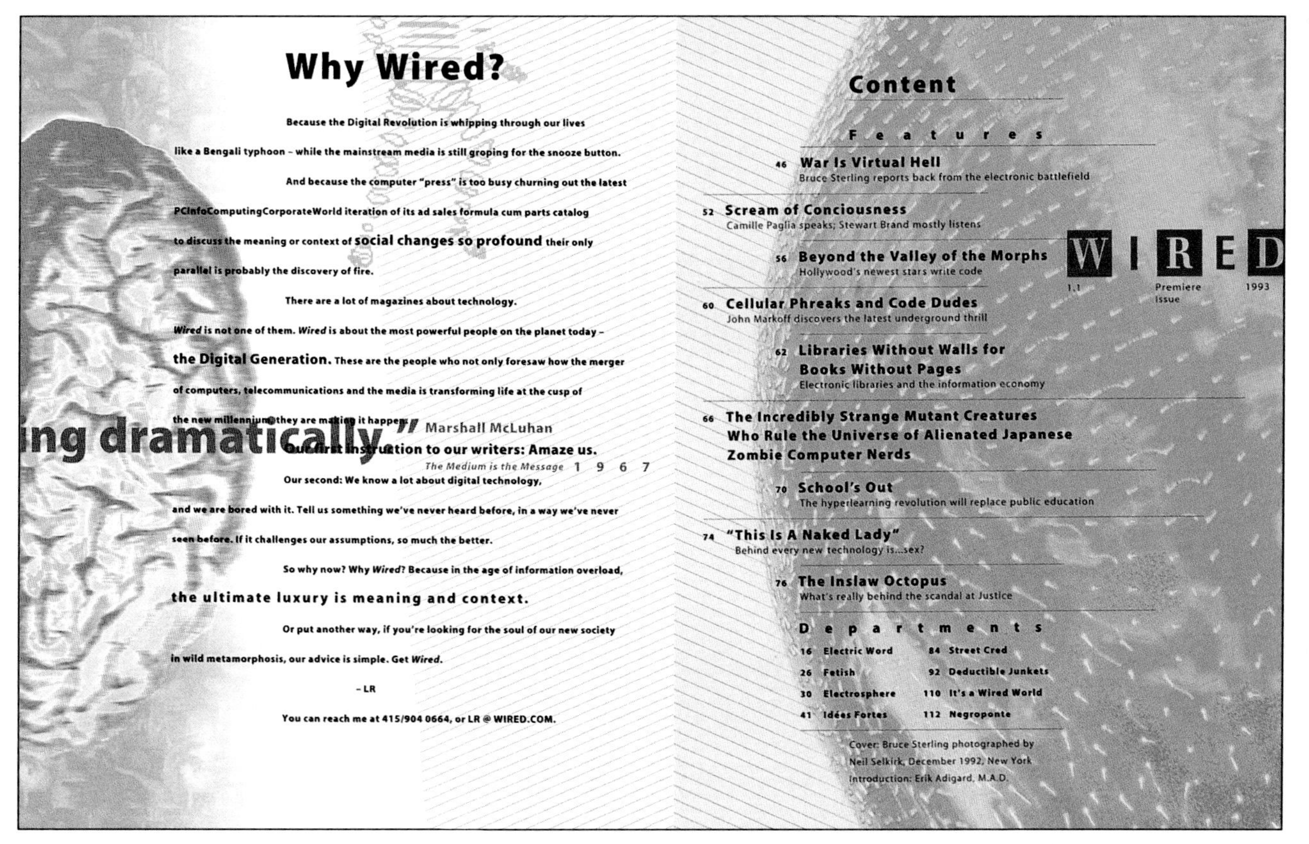

Why Wired?

Because the Digital Revolution is whipping through our lives like a Bengali typhoon – while the mainstream media is still groping for the snooze button.

And because the computer "press" is too busy churning out the latest PCInfoComputingCorporateWorld iteration of its ad sales formula cum parts catalog to discuss the meaning or context of **social changes so profound** their only parallel is probably the discovery of fire.

There are a lot of magazines about technology. *Wired* is not one of them. *Wired* is about the most powerful people on the planet today – **the Digital Generation.** These are the people who not only foresaw how the merger of computers, telecommunications and the media is transforming life at the cusp of the new millennium, they are making it happen.

Our first instruction to our writers: Amaze us.

Our second: We know a lot about digital technology, and we are bored with it. Tell us something we've never heard before, in a way we've never seen before. If it challenges our assumptions, so much the better.

So why now? Why *Wired*? Because in the age of information overload, **the ultimate luxury is meaning and context.**

Or put another way, if you're looking for the soul of our new society in wild metamorphosis, our advice is simple. Get *Wired*.

– LR

You can reach me at 415/904 0664, or LR @ WIRED.COM.

Content

F e a t u r e s

46 **War Is Virtual Hell**
Bruce Sterling reports back from the electronic battlefield

52 **Scream of Conciousness**
Camille Paglia speaks; Stewart Brand mostly listens

56 **Beyond the Valley of the Morphs**
Hollywood's newest stars write code

60 **Cellular Phreaks and Code Dudes**
John Markoff discovers the latest underground thrill

62 **Libraries Without Walls for Books Without Pages**
Electronic libraries and the information economy

66 **The Incredibly Strange Mutant Creatures Who Rule the Universe of Alienated Japanese Zombie Computer Nerds**

70 **School's Out**
The hyperlearning revolution will replace public education

74 **"This Is A Naked Lady"**
Behind every new technology is...sex?

76 **The Inslaw Octopus**
What's really behind the scandal at Justice

D e p a r t m e n t s

16 Electric Word
26 Fetish
30 Electrosphere
41 Idées Fortes
84 Street Cred
92 Deductible Junkets
110 It's a Wired World
112 Negroponte

Cover: Bruce Sterling photographed by Neil Selkirk, December 1992, New York
Introduction: Erik Adigard, M.A.D.

Wired's *statement of intent (left) was printed in its premier issue.*

was a great idea, and wanted to invest in it personally—which he did. Everything followed from there. Nicholas' interest raised the profile considerably, and his involvement facilitated subsequent investment. With seed capital secured, they were able to gather the initial team together and began work on the first issue. A San Francisco investment bank raised the additional financing required to launch the publication.

Wired magazine is now actually starting to break even a year ahead of their own business plan—which is ambitious for the industry. Most magazines don't break even for three to five years. They thought they could do it in two years, but they actually began breaking even after three or four issues.

Originally, *Wired* planned to go monthly in January 1994, but due to the tremendous reader and advertising response, *Wired* decided to publish monthly as of November 1993. With a circulation of 185,000 primarily on newsstands as of mid-summer 1993, they expect readership to increase. Their goal is to have a half million subscribers after five years.

Wired is printed on a six-color press, a design choice reflecting *Wired*'s business strategy. Since a major goal of the magazine is to place a higher value on the content, it was important to reflect that in the physical product as well. They are using more of an annual report quality as a way to affect the reader's perception of the magazine.

ELEMENTS OF THE PROCESS

John Plunkett and Barbara Kuhr are the Creative Directors for *Wired*. They are responsible for everything about the look and feel of the magazine. They say their role is to respond to the stories that are created and develop a way to get the message across visually as well as in words.

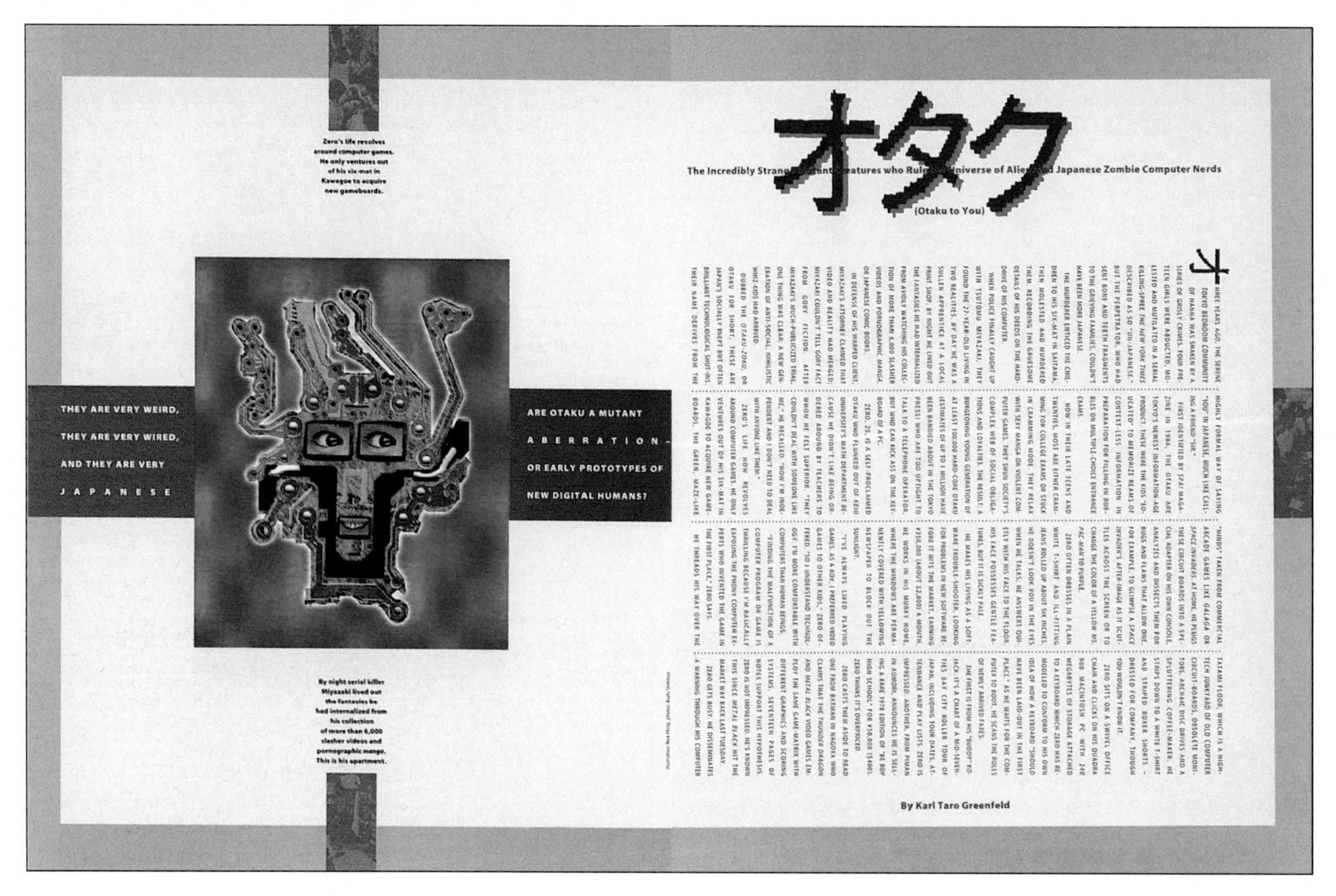

Zero's life revolves around computer games. He only ventures out of his six-mat in Kawagoe to acquire new gameboards.

オタク

The Incredibly Strang… ant …eatures who Rul… niverse of Alie… d Japanese Zombie Computer Nerds

(Otaku to You)

THEY ARE VERY WEIRD, THEY ARE VERY WIRED, AND THEY ARE VERY J A P A N E S E

ARE OTAKU A MUTANT A B E R R A T I O N – OR EARLY PROTOTYPES OF NEW DIGITAL HUMANS?

By night serial killer Miyazaki lived out the fantasies he had internalized from his collection of more than 6,000 slasher videos and pornographic manga. This is his apartment.

By Karl Taro Greenfeld

Designers John Plunkett and Barbara Kuhr challenge our preconceptions of what a magazine should look like.

Wired is the first magazine on which John and Barbara have worked. They actually saw that as one of their greatest advantages, because they came to it without any preconceptions about how it "should" be done. In the context of magazines, they're trying to signal that *Wired* is new and different, and are trying to separate themselves from the crowd. Their technique was to study what most magazines look like. John and Barbara quickly discovered that most magazines look alike—there is a particular formula most magazines follow. They made a list of what magazines do, which became their list of what they would *not* do.

For instance, they do not hold to a straight grid—they don't have preordained space into which words fill the space. According to John, part of this is a design issue, but it's more important to *Wired* to break the traditional way that magazines present information. He explains:

> *Whether it's a book or magazine, pretty much up until the last few years, the way text was presented was put in place by Guttenberg. It was based on hot metal type locked into a metal chaise to create a page.*

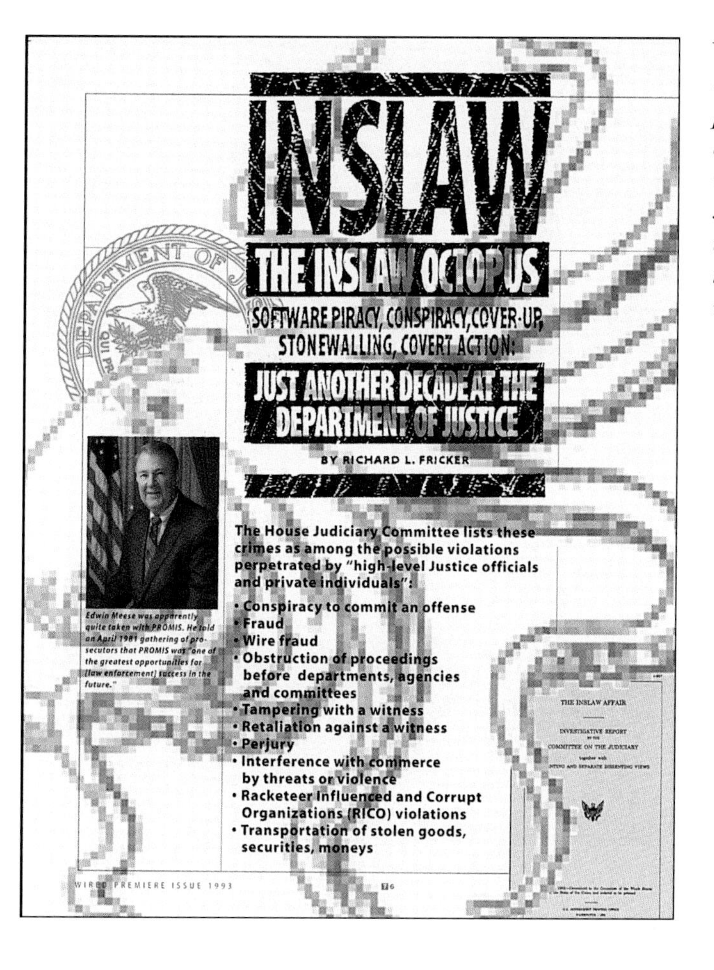

INSLAW

THE INSLAW OCTOPUS

SOFTWARE PIRACY, CONSPIRACY, COVER-UP, STONEWALLING, COVERT ACTION:

JUST ANOTHER DECADE AT THE DEPARTMENT OF JUSTICE

BY RICHARD L. FRICKER

Edwin Meese was apparently quite taken with PROMIS. He told an April 1981 gathering of prosecutors that PROMIS was "one of the greatest opportunities for [law enforcement] success in the future."

The House Judiciary Committee lists these crimes as among the possible violations perpetrated by "high-level Justice officials and private individuals":

- Conspiracy to commit an offense
- Fraud
- Wire fraud
- Obstruction of proceedings before departments, agencies and committees
- Tampering with a witness
- Retaliation against a witness
- Perjury
- Interference with commerce by threats or violence
- Racketeer Influenced and Corrupt Organizations (RICO) violations
- Transportation of stolen goods, securities, moneys

WIRED PREMIERE ISSUE 1993

Wired's electronically produced images are designed to respond to its stories and visually convey the written message.

Now, because of computers, there no longer is any hot metal and there's no reason to lock everything into a structure the way it used to be. As a consequence you see two trends in design. One is designers who go crazy with the computer and make these incredibly beautiful things to look at that are almost impossible for anyone to read. The other way that designers have been using computers is to very carefully recreate traditional hot metal standards.

Wired has been trying to find a way that explores the fluidity of the computer, since a computer allows a designer to be much more flexible.

Electronic Magazines

Wired is available on *America On-Line*. Ultimately, they expect that *Wired* will exist as an electronic medium, although they can't quite tell you what that means. Says John:

> *Everyone talks about interactive magazines and we have yet to see anything remotely intelligent in that area. We've seen a lot of really dumb, slow things that are bad imitations of magazines on screens that are difficult to read. So that's probably what the magazines of the future won't become. But eventually we do see that there'll be a shift in the weight of communication from people receiving information on paper. It will become more electronic. We intend to be there and hopefully do it well and point the way. We just don't know what that will be yet.*

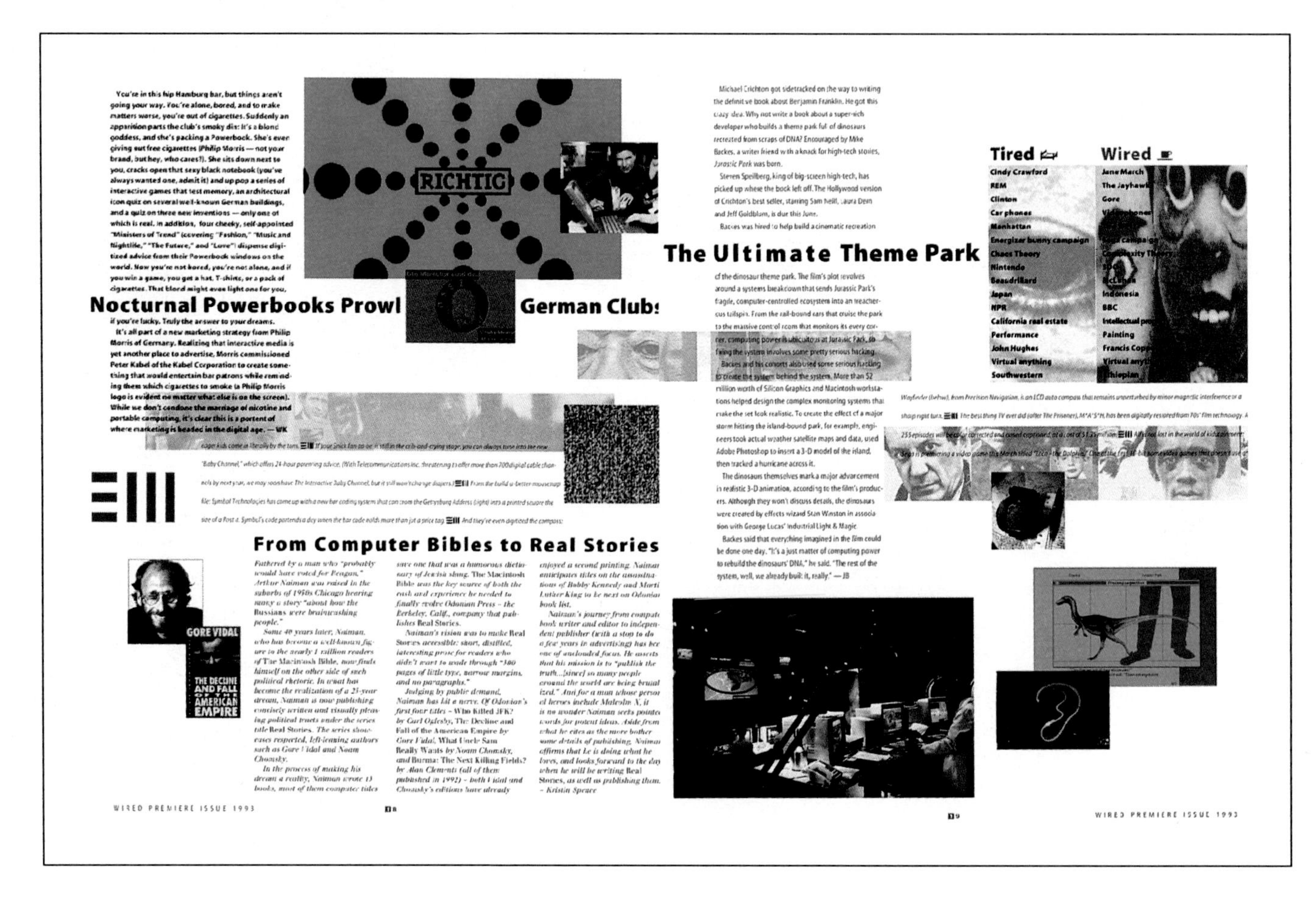

You're in this hip Hamburg bar, but things aren't going your way. You're alone, bored, and to make matters worse, you're out of cigarettes. Suddenly an apparition parts the club's smoky din: It's a blond goddess, and she's packing a Powerbook. She's even giving out free cigarettes (Philip Morris — not your brand, but hey, who cares?). She sits down next to you, cracks open that sexy black notebook (you've always wanted one, admit it) and up pop a series of interactive games that test memory, an architectural icon quiz on several well-known German buildings, and a quiz on three new inventions — only one of which is real. In addition, four cheeky, self-appointed "Ministers of Trend" (covering "Fashion," "Music and Nightlife," "The Future," and "Love") dispense digitized advice from their Powerbook windows on the world. Now you're not bored, you're not alone, and if you win a game, you get a hat, T-shirts, or a pack of cigarettes. That blond might even light one for you,

Nocturnal Powerbooks Prowl German Clubs

if you're lucky. Truly the answer to your dreams.

It's all part of a new marketing strategy from Philip Morris of Germany. Realizing that interactive media is yet another place to advertise, Morris commissioned Peter Kabel of the Kabel Corporation to create something that would entertain bar patrons while reminding them which cigarettes to smoke (a Philip Morris logo is evident no matter what else is on the screen). While we don't condone the marriage of nicotine and portable computing, it's clear this is a portent of where marketing is headed in the digital age. — WK

RICHTIG

"Baby Channel," which offers 24-hour parenting advice.

Symbol Technologies has come up with a new bar coding system that can cram the Gettysburg Address (right) into a printed square the

Symbol's code portends a day when the bar code holds more than just a price tag. And they've even digitized the compass:

From Computer Bibles to Real Stories

Fathered by a man who "probably would have voted for Reagan," Arthur Naiman was raised in the suburbs of 1950s Chicago hearing many a story "about how the Russians were brainwashing people."

Some 40 years later, Naiman, who has become a well-known figure to the nearly 1 million readers of The Macintosh Bible, *now finds himself on the other side of such political rhetoric. In what has become the realization of a 25-year dream, Naiman is now publishing concisely written and visually pleasing political tracts under the series title* Real Stories. *The series showcases respected, left-leaning authors such as Gore Vidal and Noam Chomsky.*

In the process of making his dream a reality, Naiman wrote 13 books, most of them computer titles save one that was a humorous dictionary of Jewish slang. The Macintosh Bible *was the key source of both the cash and experience he needed to finally evolve Odonian Press – the Berkeley, Calif., company that publishes* Real Stories.

Naiman's vision was to make Real Stories *accessible: short, distilled, interesting prose for readers who didn't want to wade through "300 pages of little type, narrow margins, and no paragraphs."*

Judging by public demand, Naiman has hit a nerve. Of Odonian's first four titles – Who Killed JFK? *by Carl Oglesby,* The Decline and Fall of the American Empire *by Gore Vidal,* What Uncle Sam Really Wants *by Noam Chomsky, and* Burma: The Next Killing Fields? *by Alan Clements (all of them published in 1992) – both Vidal and Chomsky's editions have already enjoyed a second printing. Naiman anticipates titles on the assassinations of Bobby Kennedy and Martin Luther King to be next on Odonian's book list.*

Naiman's journey from computer book writer and editor to independent publisher (with a stop to do a few years in advertising) has been one of unclouded focus. He asserts that his mission is to "publish the truth...[since] so many people around the world are being brutalized." And for a man whose personal heroes include Malcolm X, it is no wonder Naiman seeks pointed words for potent ideas. Aside from what he cites as the more bothersome details of publishing, Naiman affirms that he is doing what he loves, and looks forward to the day when he will be writing Real Stories, *as well as publishing them.* – Kristin Spence

GORE VIDAL
THE DECLINE AND FALL OF THE AMERICAN EMPIRE

Michael Crichton got sidetracked on the way to writing the definitive book about Benjamin Franklin. He got this crazy idea. Why not write a book about a super-rich developer who builds a theme park full of dinosaurs recreated from scraps of DNA? Encouraged by Mike Backes, a writer friend with a knack for high-tech stories, *Jurassic Park* was born.

Steven Spielberg, king of big-screen high-tech, has picked up where the book left off. The Hollywood version of Crichton's best seller, starring Sam Neill, Laura Dern and Jeff Goldblum, is due this June.

Backes was hired to help build a cinematic recreation

The Ultimate Theme Park

of the dinosaur theme park. The film's plot revolves around a systems breakdown that sends Jurassic Park's fragile, computer-controlled ecosystem into an treacherous tailspin. From the rail-bound cars that cruise the park to the massive control room that monitors its every corner, computing power is ubiquitous at Jurassic Park, so fixing the system involves some pretty serious hacking.

Backes and his consorts also used some serious hacking to create the system behind the system. More than $2 million worth of Silicon Graphics and Macintosh workstations helped design the complex monitoring systems that make the set look realistic. To create the effect of a major storm hitting the island-bound park, for example, engineers took actual weather satellite maps and data, used Adobe Photoshop to insert a 3-D model of the island, then tracked a hurricane across it.

The dinosaurs themselves mark a major advancement in realistic 3-D animation, according to the film's producers. Although they won't discuss details, the dinosaurs were created by effects wizard Stan Winston in association with George Lucas' Industrial Light & Magic.

Backes said that everything imagined in the film could be done one day. "It's a just matter of computing power to rebuild the dinosaurs' DNA," he said. "The rest of the system, well, we already built it, really." — JB

Tired	Wired
Cindy Crawford	Jane March
REM	The Jayhawk[illegible]
Clinton	Gore
Car phones	[illegible]
Manhattan	[illegible]
Energizer bunny campaign	[illegible] campaign
Chaos Theory	Complexity Theory
Nintendo	[illegible]
Beaudrillard	[illegible]
Japan	Indonesia
NPR	BBC
California real estate	Intellectual pro[illegible]
Performance	Painting
John Hughes	Francis Cop[illegible]
Virtual anything	Virtual any[illegible]
Southwestern	[illegible]thiopian

from Precision Navigation, is an LCD auto compass that remains unperturbed by minor magnetic interference or a

The best thing TV ever did (after The Prisoner), M*A*S*H, has been digitally restored from 70s' film technology.

WIRED PREMIERE ISSUE 1993 8

9 WIRED PREMIERE ISSUE 1993

The "Electric Word" section deliberately challenges the way we read, reflecting electronic rather than print communications.

They are trying to take advantage of what the computer can do and reflect what the computer can achieve, and at the same time not make it so visually overwhelming that it overwhelmes the content.

John and Barbara strive to make the content of the magazine visible. The magazine presents a lot of interesting ideas. Their approach is to take a story and visually make some of these ideas apparent to the reader in a way that compliments the story.

There's a basic contradiction about *Wired* magazine: It's a static printed medium that has existed forever, it reports on a very fluid electronic medium—one that's virtually a moving target that's impossible to catch. As a result, John and Barbara are trying to develop a vocabulary of visual metaphors to represent to the reader the electronic world on which *Wired* is actually reporting.

There is a section of the magazine, "Electric Word," which is one area where they concentrate on that message. They deliberately break through the traditional three-column, left-to-right, top-to-bottom read, and cut through that horizontally in a literal way, and have stories that are unanchored. In fact, the first time most readers see this section it is somewhat disorienting. Their hope is that the sense of being disoriented is not unlike the first

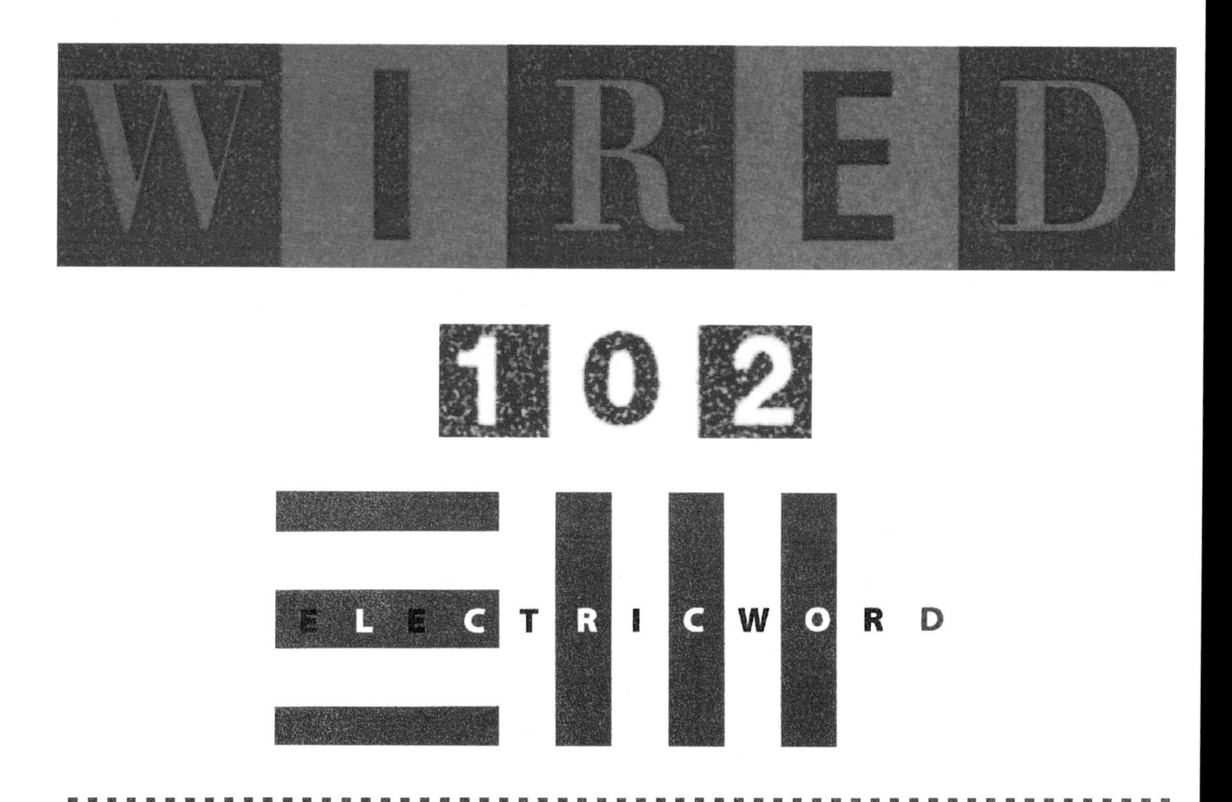

The digital nature of the Wired *logo is carried through into folios and other graphic elements.*

time someone sits down and tries to log onto a computer, walk through it's inner workings, and learn something.

The Production Cycle

Each issue of *Wired* has a 20-week cycle, from the day they begin to think about it until an issue gets into the reader's hands. It starts in a room with the publisher and the senior editorial people sitting down and talking about what they want to put in the issue. Shortly after, there's a second meeting with those people, as well as the senior design people. That's where the content, or most of it, gets decided.

During the first month, there are more meetings where stories actually are assigned; more specifically, art is assigned for those stories. That process continues into the following month when stories start coming in, and the design process

CREATING THE *WIRED* LOGO

John and Barbara created the *Wired* logo. Their main goal was to make a visual vocabulary that has to do with being computerized and digital. They designed the logo, with its three boxes, to have the effect of being on-off, black-white, zero-one, yes-no.

They selected both a modern sans-serif and a traditional serif typeface to make a point about the magazine, since at this time *Wired* views itself as somewhat of a bridge between the old world and the new world. In one sense, they're saying they don't agree that people have shorter attention spans and won't read anymore—*Wired* is, first and foremost, meant to be read. You will also see the squares of the logo reflected throughout the magazine as a "digital signature," in somewhat of a hallmark fashion.

Stars like Peter Gabriel, U2, Todd Rundgren, Prince, David Bowie, and the Residents are launching interactive Rock and Roll.

I WANT MY (DESKTOP) MTV!

"The interactive format allows people inside the work, in the interior, so they can be participants instead of observers on the exterior. That's a fundamental difference." – Peter Gabriel

(But can you dance to it?)

Fred Davis reports.

❶ Peter Gabriel's *Explora*

You're Peter Gabriel. You've been at the cutting edge of everything. You were in a supergroup. You left to go solo and earned superstar status on your own. Your videos made MTV worth watching. Your mantelpiece is crammed with Grammys and other awards. So, what's next?

For Peter Gabriel, the answer was obvious: interactive music. Gabriel's *Explora* CD-ROM, due out this fall for $49.95, includes music and videos from Gabriel's latest album *Us*; an interactive tour of his recording studio, situated on the grounds of a beautiful estate somewhere in the English countryside; and a visit to a World Organization of Music and Dance (WOMAD) festival, all with Gabriel acting as your interactive tour guide.

By letting people look at the artistic process from the inside, Gabriel hopes his work will help blur the distinction between artist and non-artist. "I hope this technology will empower people who have a sense that they have as much right and ability for self-expression as anyone who goes under the officially approved category [of artist]," he says. "In some societies, the idea that anyone alive would not be an artist is ridiculous. I think that often the first generation of new technology can be dehumanizing; but the second generation can be super-humanizing."

Interactive technology could change both the public's conception of musician as well as the economics of the music world. Gabriel isn't alone in grasping the importance of music's newest medium: Every major Hollywood studio and media conglomerate is jockeying for a stake in the latest digital opportunity. Time-Warner, Viacom, Sony, MCA, Paramount, Fox, Philips – you name it – they've all got an interactive strategy. In fact, refusing to acknowledge interactive music nowadays would be about as smart as pretending video didn't matter during the ascendency of MTV.

Not everyone is so sanguine, however. "There are real questions in my mind as to whether even talented musicians can make a compelling interactive experience," comments Jac Holzman, founder of Elektra Records, now chief technologist at Warner Music.

Indeed, a cynic might argue that interactive music is yet another marketing trick foisted upon the consuming masses by aging rock stars eager to extend their popular lifespan. "MTV was essentially a new marketing methodology," Holzman says.

"The life of a rock-and-roll musician used to be short," he adds. "Now it can be 20 to 25 years."

Holzman sees more renowned stars like Gabriel, Bowie, and others raising awareness of interactive medium's potential. But the artists who will truly define this new form have yet to be discovered.

"It's like the growth of the record industry as it came out of the '50s and '60s," says Holzman, who began his career when records still spun at 78 rpm. "The energy was with the independent labels." Inexpensive tools like Apple's QuickTime and Adobe Premiere will enable a whole new generation of artists to explore this medium, he adds.

Until then, veteran stars like Peter Gabriel, Billy Idol, David Bowie, and scores of others are working with computer technologists to add an interactive dimension to the music they've already written. Others, including Todd Rundgren, Thomas Dolby, and Cindy Baron, are creating music from the ground up using the new interactive technologies. Whether they create interactive music by integrating old tunes into new formats or by writing new music, both approaches yield what can only be described as an altered listening experience.

WIRED JULY/AUGUST 1993 4

Bamboo Harp

MADAGASCAR

"There are real questions in my mind as to whether even talented musicians can make a compelling interactive experience." – Jac Holzman, founder of Elektra Records, now chief technologist at Warner Music.

The Rise and Fall of the Haight-Ashbury

❷ Tony Bové's *The Rise and Fall of the Haight Ashbury*

❸ Interactive Record's *So You Want to Be a Rock and Roll Star*

The first wave of multimedia music products falls into three main categories: ❶ interactive musical compositions, which let you make your own music out of pre-constructed parts, or jam along; ❷ interactive rockumentaries, which bring you into the history of the creative musical process; and ❸ interactive multimedia experiences, which combine music, art, and even a sense of gameplay.

New releases from artists like Gabriel will let you do more than merely listen to your music. You'll be able to play along with the band, even if you're a tone-deaf klutz. You'll have the chance to create your own music videos, identify your emotional responses to certain kinds of music, and even explore a virtual musical world. And we're not talking some distant future – interactive rock and roll is happening right now.

Technologists and Artists

Today, creating interactive CD-ROMs requires both musical and technical knowledge. Some CD-ROMs, like *The Freak Show*, under development by the San Francisco-based performance-rock group The Residents, consume the talents of computer programmers, artists, designers, and computer-literate musicians. Musicians like Bowie, whose demo CD-ROM was produced by interactive wizard Ty Roberts, and Gabriel, who enlisted the help of technogeek Steve Nelson of Brilliant Media to produce the *Explora* CD-ROM, are working with the best technologists around.

The Los Angeles-based hard-rock group Mötley Crüe worked with programmer Tim Byers to produce a CD-ROM that will give you instant access to their hit song *Girls! Girls! Girls!* Other efforts, such as the *No World Order* CD-ROM by Todd Rundgren, is the sole creation of the renaissance man himself.

There's even a non-musician, Tony Bové, an interactive chronicler and historian, whose CD-ROM-based documentary-in-progress about the Haight-Ashbury days is based on recorded music, art, poetry, and news footage.

Make Your Own Music

The first wave of interactive music falls into three main categories: interactive musical compositions, which let you make your own music out of pre-constructed parts or let you jam along with the music; interactive rockumentaries, which bring you into the history of the creative musical process; and interactive experiences, which combine music, art, and even a sense of gameplay.

Interactive musical compositions include works with the primary focus on audio. The visual interfaces to these products let you set the mood of the music that's playing, remix or rearrange the various musical parts in the composition, or let you play along as one of the band members. "Any musical performance is actually a script of musical events: verses, choruses, solo sections, and other subjectively named musical events," says Todd Rundgren. "A traditional CD is essentially a list of these musical events.

"On my interactive CD, there will be three pre-defined scripts of musical events available. At the simplest level, it's like getting three albums in one, and you can choose the one you like best. If you want more, you can have the system create a custom script for you. You can control the tempo or mood. Or maybe you want to hear a certain kind of mix, like a very thick mix or a thin mix or a mix with no vocals."

Experimentation with mood and emotion is a major theme in the interactive music scene. Thomas Dolby describes how he designed a sound installation for a sculpture exhibit in an art gallery: "The nice thing about music is that when you add an element to a piece of music, your whole perception of it changes. You take a chord sequence played with a mellow choir sound in C major, and then somewhere else in the room you place a low throb or drone in A minor. As you move

5 WIRED JULY/AUGUST 1993

Color is an important element to creative directors Barbara and John, as shown in this spread.

begins. Physically, the writers are submitting their work electronically on disk or via modem. The magazine gets very little copy in as hard copy, since words are definitely the easiest thing to send electronically, and almost anybody can do it. There are still a few people working with typewriters, but not many.

Once the first round of editing is complete, the stories are placed into QuarkXPress documents to be laid out. The text is formatted roughly, and the QuarkXPress files are posted in a specific folder on a network server. *Wired* has a folder structure that replaces in and out baskets. Once a file is placed in a certain folder, the editorial and production staff knows it's time to perform a particular action.

They also have an analog system in which a paper folder exists for each digital one. On this paper trail, each person checks off what he or she has done. When a milestone is achieved, a staff member signs off on it and hands the paper folder to the next person.

The roughly composed pages are returned to editorial for copy fitting, then routed to the design department for a final layout, where art is paced, and back to editorial for

S T R E E T   C R E D

New Techno Rag: *Black Ice*

Motivated, as it were, "in part by *Mondo 2000's* decline in editorial content" a group of writers, graphic designers, and photographers in London banded together to start *Black Ice* magazine, a new member in the rising chorus of voices seeking to explore and articulate where technology is headed.

We've only seen the first issue (the second was printed after deadline). Tasty treats include an interview with Jon Cline and Mark Pellington – creators of the visionary (albeit canceled) 1991 MTV series *Buzz*, a naughty *Star Trek TNG* comics spoof, and a look at the inchoate world of Japanese junk food. Also inside is perhaps England's first take on Generation X, clever instructions on how to make a nuclear device, a lot of VR news (much vapor, some meat), and a semi-academic reading of movies-as-drugs. Unfortunate is the fact that interviews are exclusively Q & A – with precious little context, interpretation, or spin from the interviewer. Best quote: "*Black Ice* will develop a fully-functioning and integrated tele-dildonic system for anyone who will put up the necessary capital." If you're virtually horny and have some cash to blow, note info below. Recommended. – *Will Kreth* •

Black Ice, $32 for four issues, Titan Distributors (UK): +44 (071) 538 8300.

Internet Vending Machines

In the mid-1970s some brilliant student at Carnegie Mellon University got tired of trudging down many flights of stairs from his computer terminal to get a soda from a basement vending machine. By some law of the universe, the machine was always out of his favorite variety whenever he was most thirsty. So he hacked up a network connection to the vending contraption. Now, before he made the long trip down, he would e-mail it to see which sodas were available.

The Coke machine on the Internet became legendary, and it wasn't long before students a thousand miles away were querying it, just on the principle that it was there. This spawned other Coke machines hidden here and there on the Net. The obvious next step: e-mail the vending machine and have the Coke delivered to your room.

While we all wait for that utopian day, other types of vending machines are popping up on the Net. These dispense information. The

idea behind the Net-wide service of FTP is that if you want some information, all you have to do is go into the machine somewhere on the Net that stores it and copy it into your machine. What could be easier?

Turns out that for those who have trouble navigating the unsigned labyrinth of the Net, a vending machine is easier. You simply stay where you are and e-mail a message to an information vending machine. It dispenses what you want back in minutes. Right now Net vending machines are free, but some time in the near future some will accept a form of electronic money for commercial material.

You can get an idea of how all this works by sending a message, any message, to info.new.technology@ieee.org. Your message goes to the vending machine at the IEEE (the electrical engineer's association) which holds its technical reports on emerging technologies. What you'll get in return the first time is a voicemail kind of menu that says "if you want this file, mail to address #2, if you want that file, mail to #5." Or, you can jump directly to a final choice and mail to info.new.technology.sit@ieee.org. In a few minutes (depending on traffic) the Internet vending machine should deposit a report-in-progress on the social implications of virtual reality in your mailbox. – *Kevin Kelly* •

Fishy-Soisse

Say the person you're dating likes to go to the symphony. Classical music bores you to tears, but you'd do anything for this affair. Well, in a few short hours, you can start talking fugues and fortissimos just like a real music critic poseur. And probably be able to start backing off that pre-concert No-Döz.

The Voyager Company of Santa Monica, California has released a classical music series on CD-ROM for Apple computers (PC versions will appear eventually). The idea is to use all the resources possible on CD-ROM to open up works of art that would otherwise be impenetrable. A few hours with Schubert's *The Trout* quintet is quite enough to make the scales fall from anyone's eyes. One not only learns how to listen to that piece but learns how to hear any piece of classical music in a whole different – and much fuller – way. – *Seth Chandler* •

Trout CD-ROM: $59. The Voyager Company: +1 (310) 451 1383.

WIRED JULY/AUGUST 1993 100

Reflective art or photography is usually placed for position only, then scanned at a high resolution and placed by their print house.

proofreading. The art is either electronic illustrations produced in-house or by a contractor, or it is reflective art from a contractor or a photograph placed for position only (FPO).

Working with Images

Wired usually has its printer scan larger color subjects on a DS America 608 drum scanner directly into the Scitex prepress system. The printer then returns low-resolution scans to the magazine, which are placed into QuarkXPress files. When the QuarkXPress file reaches the printer, they perform an automatic picture replacement (APR).

The benefit of this system is that *Wired* still controls where images go, and they don't have to rely on someone else to eye their proofs

BRINGING COLOR SEPARATIONS IN-HOUSE

Eugene Mosier, Production Art Director, is exploring the possibility of bringing more of the color separation work in-house. Currently, their printer is handling their prepress work as part of a package deal. As a result, their prepress costs are so low it doesn't really make a lot of sense for them to bring the larger color subjects in-house.

Because of the processing power needed on a Mac, they would be making a rather incredible capital investment in terms of buying a lot of RAM, hard disk space, and fast computers. That would pay for itself pretty quickly, but the next problem is that no matter how much equipment there is, it really comes down to the knowledge and experience of the operator. Says Eugene:

> *The fact of the matter is that I sort of know what I'm doing, but compared to somebody who does Scitex work, day in and day out, five days a week, I know very little.*
>
> *There are people at our printer who, with their eyes closed, without even looking at the image, can probably separate it better than I can on the Mac because they can look at the numbers and understand what those values are. A Scitex operator generally works off numbers more than what they're seeing. They just use the visual feedback for information purposes.*

IBM gets credibility.
***T-2's* Jim Cameron gets $20 million.**

Matt Rothman gets the story behind Digital Domain, Hollywood's newest digital studio.

The scene is pure Hollywood. At one of Beverly Hills's toniest hotels, journalists mill around as pasta in cream sauce and spicy chicken legs bubble in catering trays. Perched in a director's chair in black leather pants and a chartreuse jacket is IBM's Lucie Fjeldstad; beside her is the more casually dressed Jim Cameron, director of several blockbuster movies.

It's not one of a hundred movie release parties. Instead, it's more of a coming-out party for this town's latest – and oddest – couple. Fjeldstad, IBM's general manager and vice president for multimedia, is now a partner with

DIGITAL DEAL

Cameron, who's best known as writer and director of the two *Terminator* mega-pics that catapulted Arnold Schwarzenegger into Hollywood's box office stratosphere.

Also adorning a specially constructed stage is Stan Winston, creator of Arnold's metallic prosthetics in *Terminator 2: Judgement Day (T-2)*, and Scott Ross, former head of George Lucas's premier special-effects house, Industrial Light & Magic (ILM). Together, this formidable pool of talent has started a new computer-based effects company they call Digital Domain.

This schmooze-debut last February has everyone wondering what would draw IBM, the financially ailing computer behemoth, into such a seemingly out-of-the-way and risky business? For a corporation that's made its reputation serving other corporations, making movies is a surprising interest.

Fielding questions from her director's chair, Fjeldstad confidently assures everyone that her new venture makes sense. Reportedly at her say-so, Big Blue plunked down $20 million on Digital Domain, with the hope that she'll make back that and much more in the next five years.

Maybe so. Digital Domain, says a smiling Fjeldstad, is where Hollywood is *really* going to go digital. Already, the entire movie-making process, from scriptwriting to editing, is going though a gut-wrenching transformation as jarring as when the talkies appeared. What was once a paper and photo-chemical process is now being invaded by people wielding PowerBooks and Silicon Graphics Irises. While movies are still shot on film, increasing portions are now directly transferred to disk drives. There, eye-popping computer-generated effects are added. Then the entire thing is edited down on the computer screen. The finished film leaves the digital realm only for processing at the photo lab, where thousands of prints are made – untouched by human hands and damn near the same quality as the original.

What IBM wants is a piece of each step in this new process. Digital Domain's aim is to roll out a host of new software applications for the special-effects business and beyond. IBM will have first crack at the software, plus it will own half of any characters and stories Cameron and Winston develop with the company – from futuristic warriors to friendly aliens.

"We can start with a special-effects, digital-production studio and make money on characters and software," says Fjeldstad, who will remain at IBM and serve as a director of Digital Domain. "That will allow us to build the tools for Digital Domain in the future. We're going to lead the next generation of computer platforms and applications."

Matt Rothman covers technology for Daily Variety *in Hollywood, and has been reporting on computers and multimedia for five years from the US and Japan.*

3 WIRED JULY/AUGUST 1993

All digitally created art is placed in QuarkXPress files, which are later ripped through a Scitex high-end prepress system.

and place the FPO images. If they set up their files correctly, images almost always appear in the right place. The system isn't perfect, sometimes a file is misplaced, but the error is usually so obviously wrong that it's easily detected.

The images that *Wired* electronically provides for the printer are either in encapsulated PostScript (EPS) or native Scitex (CT) format created from a Photoshop file.

Wired is using a new digital color proofing process—Kodak Approval. It allows the printer to make a proof on the correct paper stock that's remarkably close to what is actually printed on press. But because it's digital, they're able to make these proofs before they make the film and plates, making it a relatively inexpensive way to proof color—a quicker, faster, cheaper solution. It's also a plus for *Wired* because it's as digital a technique as is available, and they are trying to be right on the cutting edge of what can be done digitally.

Composed QuarkXPress pages are finally converted to PostScript and ripped via the Scitex Universal Gateway to the Scitex system's Micro Assembler, where the EPS images are automatically replaced with the color-corrected, high-resolution scans.

Final film is plotted on an Eray, and the magazine is printed in six colors on a Harris-Heidelberg M-1000 web press.

For the most part, *Wired* has had very little trouble producing a fully electronic magazine, although there are occasional rough spots. Eugene Mosier, Production Art Director for *Wired,* says that most of the problems they deal with have to do with the fact that the publishing industry—a huge, very expensive industry—is still reeling from desktop publishing. The industry is trying hard to keep up, but a lot of things are moving faster than the market can handle.

EQUIPMENT

- Macintosh II family of computers
- Radius Precision 2-page display systems, Precision Color Calibrators, and Rocket accelerators
- Dataproducts LZR 1560 laser printer
- Hewlett Packard Scanjet IIc
- APS SyQuest drives
- Micronet DAT backup and gigabyte drives
- Farallon Ethernet and LocalTalk networking
- QuarkXPress
- Adobe Illustrator
- Adobe Photoshop
- Kai's Power Tools Photoshop plug-ins
- Fractal Design Painter
- Altsys Fontographer
- Adobe Dimensions, Alias Sketch, Electric Image Animation System, and Specular Infini-D 3-D illustration programs
- Aldus Fetch, Second Glance ScanTastic PS Photoshop Plug-in, and Equilibrium Technologies DeBabelizer graphics support packages
- QuickMail, Star Nine, and Microphone networking support

CONVERSATIONS

JANE METCALFE
President and Co-Founder of *Wired*

Wired communicates how technology is changing our lives. It's more about the impact, where things are going, what the trends are, and how to get some perspective on it. There are quite a few magazines out there (and I would classify the bulk of the computer magazines in this category) that focus on explaining technology to people. They say, "Here's the latest product from company X, Y, or Z" and "Here's how to use it. Here's how to be more effective with that product." This is *Wired*'s mission.

I'm fond of saying that as the technology changes us, the effect is not just on business people, but on people with lives and weekends and families and outside interests.

There's a wonderful quote in our media kit from a man who's director of advanced technology at Lotus. He basically says he works to develop all this stuff because he's trying to figure out how it's going to make his life easier. Bill Atkinson, who's now the chairman of General Magic, will get up on stage and talk about why he's created, or why he's interested in seeing this technology development, and then his entire demonstration will be about sending messages to his kids and his wife throughout the day reminding them that he loves them. This is what's really amazing. To take our own example a bit closer to home, technology enables John Plunkett to live in Park City, Utah, and work with us remotely from various locations and still stay in touch with the office.

JOHN PLUNKETT
Creative Director

John describes himself as a "general practitioner" of design. He and his partner, Barbara Kuhr, are currently designing an exhibition for Carnegie Hall, the film trailer and graphics for the Sundance Film Festival, and* Wired *magazine. Plunkett and Kuhr left New York three years ago to open their office in Park City, Utah. They realized the technology would allow consultants to live wherever they want. They fly to New York or the West Coast every few weeks for the same meetings they would have had if they were still on West 19th Street. Everything else is done by fax, phone, and modem.

We're trying to invent something new with *Wired* in a number of different senses. In one sense, we're trying to go back to traditional journalism, meaning when the content of the story was highly valued. We have a sense that most magazines these days have become a little like TV networks, in that the content is seen as filler to aid an advertising vehicle. We wanted to go back to the way magazines used to be, whether it was *Esquire* in the '60s or Henry Luce with *Life,* where people came to a magazine to learn something, and they really valued it. We felt that if we did that job well, we could place a higher value on it. We charge $5 for an issue rather that the $2 or $3 that's charged for most monthly magazines. But people seem to be responding to that idea. So in that sense, it's going back to some traditional roots of magazine reporting.

I think what's makes it most interesting, though, is that we're applying this approach to report on what we believe is the most significant change happening today: the convergence of communications and digital technology with entertainment and information—basically a whole new way for people to receive information. The role of the magazine is really to report on technology and its effects on people, but to focus less on hardware and software, and more on the people, almost from an anthropological point of view.

We're looking at the technology as a cultural phenomena, not as nuts and bolts. We would have liked to have started this magazine, had we been able to get it financed, a year or two earlier. In retrospect, that might have been too early. It seems, by coincidence, we arrived at the right moment when it was beginning to get into the popular consciousness that something is going on here.

What we're trying to do with *Wired* is take the popular conception of technology, which is basically a mental picture of some nerd in the basement with a computer, and turn that around. We'd like to point out to people that in fact, everyone to some degree is using computers or being used by them, and it's no longer a nerd in the basement. In fact it's some very cool people, and may even be the most interesting thing going on!

EUGENE MOSIER
Production Art Director

Eugene has a degree in fine arts (sculpture), and has worked in the video production industry. After leaving video post production, he entered the publishing field as a support technician for computer networking and production. He's the person who tells the designers whether or not they can implement their ideas on a computer, and, if they still want to do it, he coordinates how it can be done traditionally. He also does color separation for the non-critical color in the magazine. Eugene heard about* Wired *magazine through a friend before the dummy issue was created, and at a party introduced himself to the magazine's founders, Jane Metcalfe and Louis Rossetto. He's been working with them ever since.

Most service bureaus don't do a lot of prepress. The way things have worked out already is that the majority of service bureaus have limited their responsibility to making sure their film density is correct. If your color doesn't look right, that's your problem.

Now, prepress houses are a different matter. They're either adopting the new technology and reducing their costs to match, or they're going out of business, or changing to a different line of business. I think we're not going to see too much stripping of film in a few years. It's not even a revolution any more. It's sort of inevitable. But in terms of what a magazine publisher would do, it is still relatively unusual for the publisher of the magazine to be doing their own prepress in-house. And that'll become more and more common. In the same way that type specing and paste-up no longer really exists, all these jobs that used to be separate people doing all these things are being condensed into the hands of a few people who don't necessarily know what they're doing.

Color separation is going to have that happen, too. And you're going to see a lot of ugly color. If you don't know what you're doing, it doesn't matter what equipment you have. You can make something ugly really quickly. But the fact of the matter is that most people don't really notice ugly color. Or they do, but they don't recognize that's what they're noticing. They say 'That looks cheap," and they don't know why it looks cheap.

So it's just a matter of people learning. Some people will learn how to do it right. When desktop publishing first started there was a lot of ugly type, and there still is, but it's getting better. I already am seeing a lot of ugly color. There are more and more ads in trade publications that are being produced desktop. There are more and more magazines that are doing desktop color and you can tell by looking at them. Including ours. I can tell by looking at our color. Not only because I did it, but I can see a quality difference between what I separated and what our printer separated. A lot of that has to do with scanners. They've got a quarter million dollar drum scanner and we're scanning a lot of stuff on a $1,000 HP Flatbed scanner. Garbage in, garbage out is one of the principle computer paradigms.

SUMMARY

The scope of *Wired* magazine—not simply its digital production, but its coverage of the digital generation—brings to light that we are becoming increasingly dependent on technology. A lot of our culture is based not on primary technology, such as internal combustion, but on secondary technology, such as using computers to do very basic tasks in our society. Furthermore, those computers themselves cannot be designed without the aids of other computers. In a sense, we are becoming more dependent on goods, services, and an economy that is increasingly controlled by a smaller and smaller number of people. Has this created a situation in which we're becoming less autonomous and really dependent upon major corporations and a global economy? Or are we becoming more independent due to our ability to access an unthinkable amount of information, communicate with millions of others, work from remote locations, and generate products limited only by our imaginations?

CHAPTER 3

NEWSPAPER

This chapter gives you a profile of the fast-paced world of newspaper production. Learn how *USA TODAY* produces a daily four-color newspaper and how it uses communications technology for national delivery.

USA TODAY

USA TODAY®

VIA SATELLITE THE NATION'S NEWSPAPER SECTION D

USA TODAY Life

WORKING WOMEN WITH KIDS

LIFELINE

REAL Fashion for Fall

Quick meals

VIA SATELLITE THE NATION'S NEWSPAPER SECTION C

USA TODAY Sports

SUPER 25 BASEBALL

SPORTSLINE

SIGNING FOR DOLLARS

Gibson vows he'll be ready

Valenzuela's outing good; Sheffield survives scare

VIA SATELLITE THE NATION'S NEWSPAPER SECTION B

USA TODAY Money

SPORT CAR LEGENDS

THURSDAY

MONEYLINE

S&L deposit hemorrhage worsened in Jan.

VIA SATELLITE THE NATION'S NEWSPAPER 50 CENTS

NCAA SWEET 16

USA TODAY

NO. 1 IN THE USA...5.3 MILLION READERS EVERY DAY

FALL'S NEW TV SHOWS

Rose probed; he's not talking

NEWSLINE

Broadcast news' stars

Inflation fear puts Wall St. to test today

Gun ban sides sign new troops

Photo courtesy *USA TODAY.*

USA TODAY is a four-color daily newspaper that is distributed nationally.

INTRODUCTION

USA TODAY is a four-color daily, national newspaper, that is produced with a combination of some traditional methods and some high tech methods. The system developed at *USA TODAY* uses almost every production system available today, from Macintosh desktop publishing stations, to Scitex color systems, to dye sublimation printers, and even a four-color proofing press.

In addition to its national daily product, *USA TODAY* produces several special bonus sections, an international edition, several other publications (*Baseball Weekly, Kids Today,* and *USA Weekend*), as well as associated advertising and promotional work.

SL: RATES10C
ED: 01 PA: 01a LEN: Y008.67/0065
FR: CMPDON-WLD;09/09,13:26
NOTE: VER: 02
09-SEP-93 13:26:57

PAGE 1A ☛ WORLD

Germany's rate cuts win cheers

The German Bundesbank, announcing a long-awaited relaxation of monetary policy, cut interest rates Thursday in a move that could help Europe emerge from an economic slump.
The central bank, which six weeks ago plunged Europe into

The text portion of USA TODAY *is generated in galley form on an Atex system.*

ELEMENTS OF THE PROCESS

The *USA TODAY* production system has two branches—text and graphics—since the technologies have evolved that way. Text systems have been very good at processing text, picture systems have been very good at processing pictures. Traditionally, text systems have not been very good at processing pictures and picture systems have not been very good at processing text.

The Text System

There are about 500 to 600 Atex terminals at *USA TODAY*. *USA TODAY* also has Newslayout, a very primitive pagination package from Atex, which allows text to come out somewhat paginated or completely paginated, so all text is output from the Atex system either in galley form or semi-paginated form. If *USA TODAY* wants to change to another type of system—such as a PC-based system—it

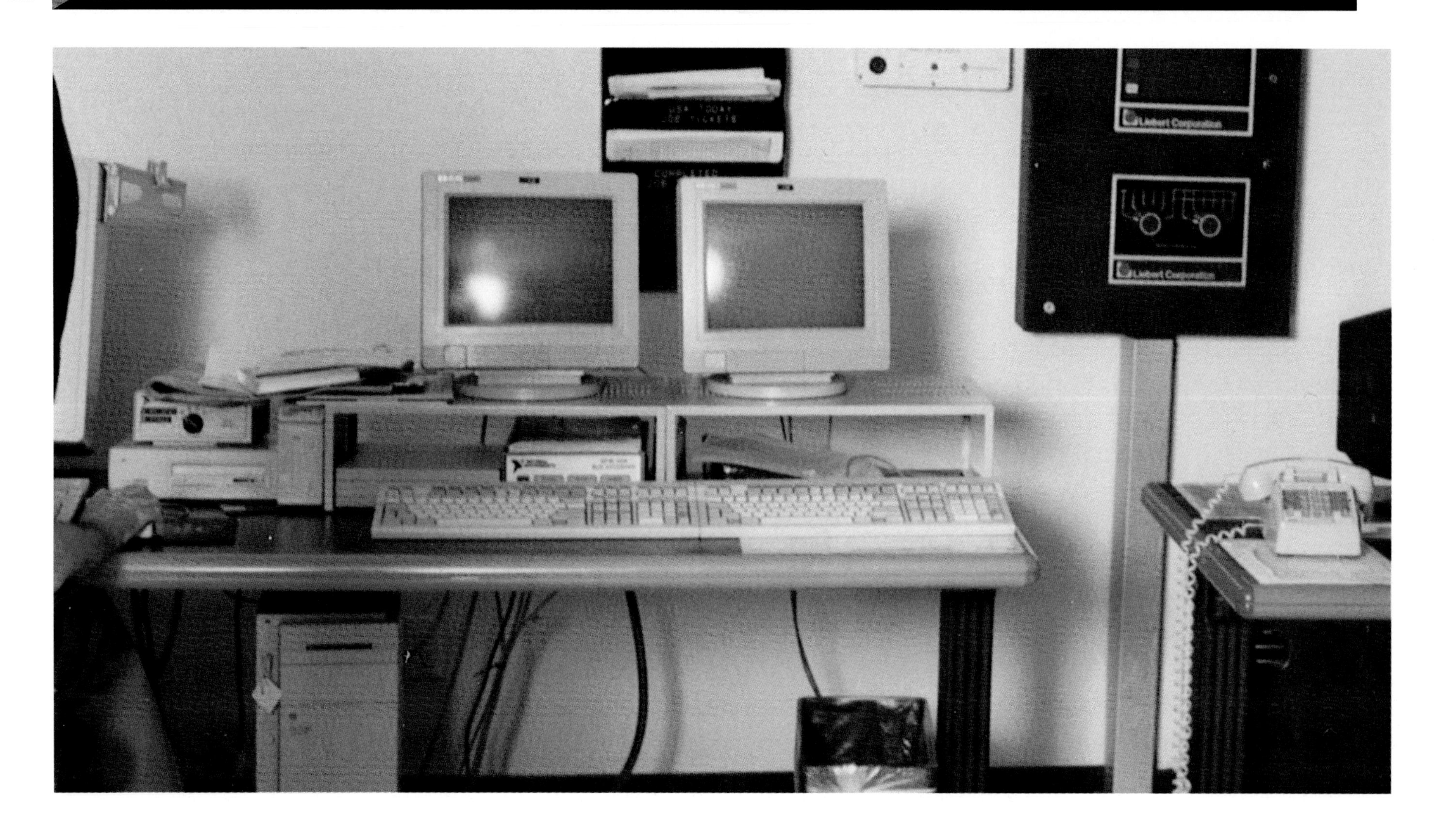

would be very expensive because the Atex is such a core part of its operation.

Atex is run on DEC's old J-11 computers. They have been around for a long time and are built for a single purpose. The software is highly developed for what it does, but it doesn't go beyond that. Because the Atex system is used so heavily for word processing—for an editorial purpose—and has good annotation capabilities, it's very useful for a number of people working on a story. A number of people can look at a story simultaneously, and they don't get a number of versions of the same piece existing on this system, which can be a problem on Macs and IBM compatibles. It's very useful for the editorial process, but it doesn't interface well with the other systems at *USA TODAY*. The next step for *USA TODAY* is to figure out what they need to do to utilize the Atex system that exists, get it to interface better with other systems, or to bring in something entirely new that allows them to do it better.

All color information is placed and separated via the Scitex system.

Photo: Robert Edward Sherbow

Because text is output in galley form, *USA TODAY* ends up making line corrections for any last-minute changes, misspellings, or errors. This is standard operation for newspapers in general, because accuracy is an absolute primary consideration, and there are always small corrections right up to the last minute.

Once galleys are generated, they are laid onto mechanicals, which is actually an advantage. The mechanicals are available for view by a number of people at once in the *USA TODAY* composing room, and a number of different people in the organization need to check these pages. This results in the ability to have a number of eyes on a given page at any time, thus the paper can be reviewed simultaneously by a number of people, not just section-by-section, one person at a time.

The *USA TODAY* staff has found this a great way to detect errors, which is a significant consideration for them. And, once a mistake is detected, the error-correction cycle is very short—it's simply a mat-

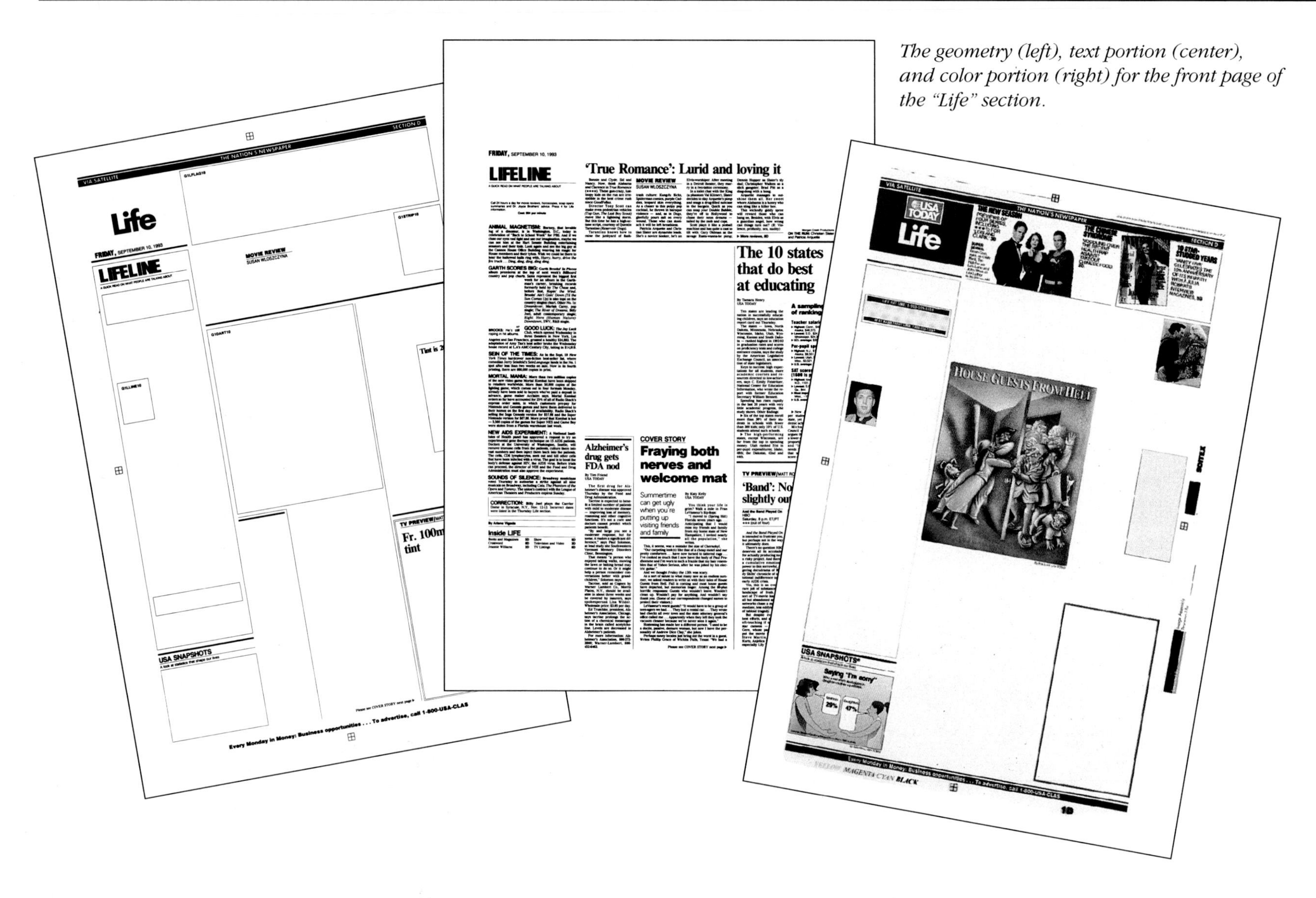

The geometry (left), text portion (center), and color portion (right) for the front page of the "Life" section.

ter of cutting it out and pasting a new piece in, which is very difficult to duplicate on an electronic system at this time. This is one of the reasons they have chosen to keep this system for as long as they have. In some cases, such as late-breaking sports scores, they even use laser proofs to place late items because they are very interested in saving one or two minutes. It's very important to get the newspaper on all four corners of the country every Monday through Friday, which they have done since its creation in 1982.

All of the page designs originated with Atex through the Newslayout program. To get that geometry into the Scitex system, they had to come out with a paper print, then manually redigitize it on the Scitex system—at this time a very

VIA SATELLITE THE NATION'S NEWSPAPER SECTION D

USA TODAY

Life

FRIDAY, SEPTEMBER 10, 1993

THE NEW SEASON

THE CHINESE SYNDROME

10 STAR-STUDDED YEARS

LIFELINE

'True Romance': Lurid and loving it

MOVIE REVIEW

HOUSE GUESTS FROM HELL

The 10 states that do best at educating

A sampling of rankings

Alzheimer's drug gets FDA nod

COVER STORY

Fraying both nerves and welcome mat

Summertime can get ugly when you're putting up visiting friends and family

TV PREVIEW

'Band': Noble, but slightly out of sync

USA SNAPSHOTS®

Saying "I'm sorry"

29% 47%

Every Monday in Money: Business opportunities ... To advertise, call 1-800-USA-CLAS

YELLOW MAGENTA CYAN BLACK

The composite image of the three portions shown on the opposite page.

inefficient process, since cost-effective interfaces have not yet hit the market.

The Graphics System

Over 20 staff members a day work *USA TODAY*'s electronic imaging department to produce the newspaper—including scanner operators, assembly operators on Scitex systems, desktop publishing staff, black-and-white imagery specialists, a support technician, supervisors, and retouching specialists.

HOW FAR IS A TOTALLY DIGITAL PRODUCT?

Rob explains that it is probably easiest to have digital production when you are talking about the editorial product, which can easily be done electronically. He remarks, however, that the largest problem with producing the entire newspaper electronically is with advertising, since most ads do not exist in an electronic form—they exist as film, and arrive at the newspaper as film:

> *It's probably easiest when you think about the editorial product. That can be done entirely electronically. It can be done. Now, whether you want to do it or not is another question. The problem with taking the whole thing electronic is the advertising, because a lot of these materials do not exist in electronic form. Probably 98% of the materials come in to us as film or Velox type paper. Something physical that you can hold in your hand. Now once you have that, if you have an electronic paper, I've got to take the hard copy materials and put that into electronic form. Then you do have an electronic paper. Now that's a process that we are not sure how to address or if we would even want to address it.*

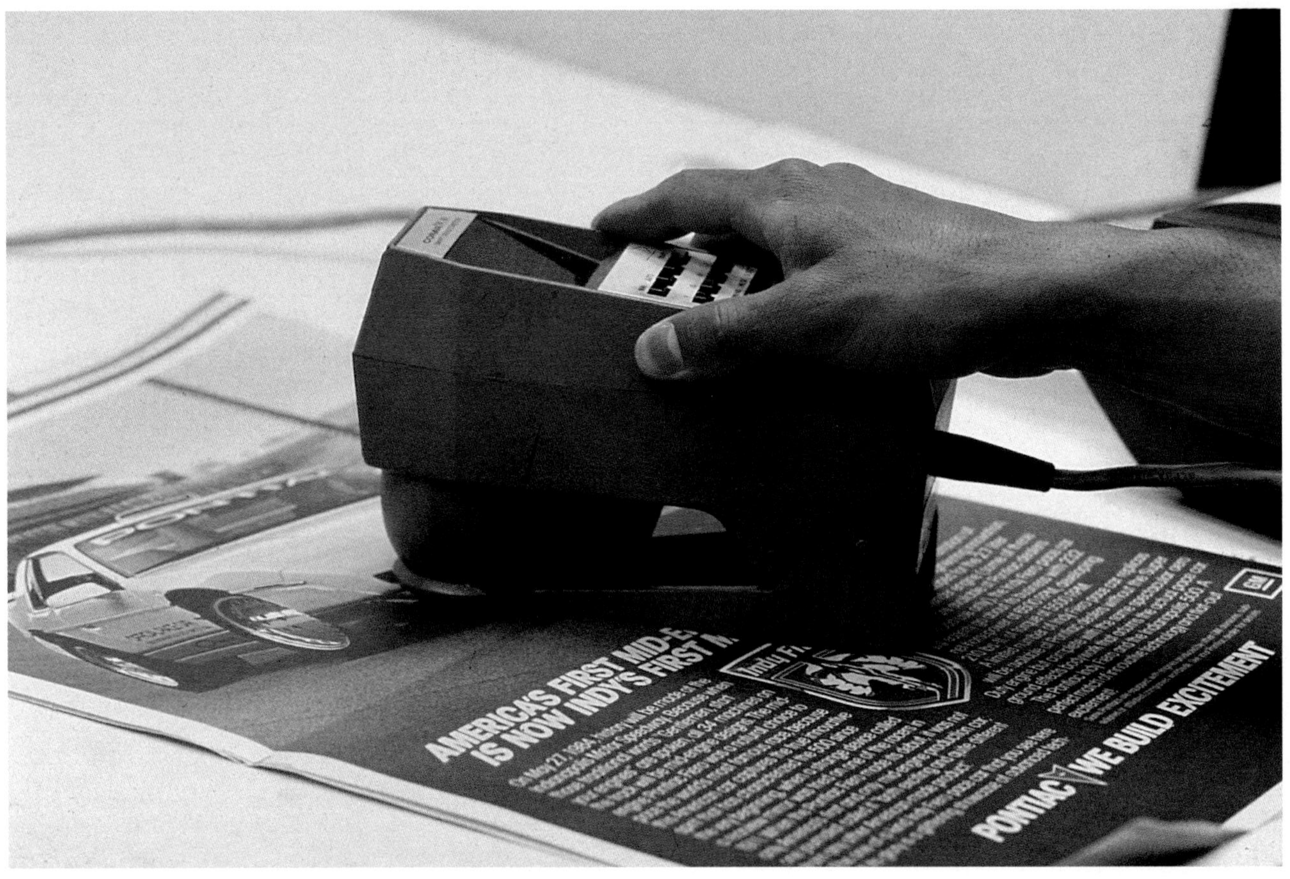

Photo courtesy *USA TODAY.*

A densitometer is used to check color on ads and photos.

Prior to installing digital production equipment, the Production department had an AP Darkroom, which was based on a Micro VAX or Mini VAX computer. It was an analog transmission system that had some capability for resampling color images. Although they could not do a great deal of work on this system, they used it experimentally until they implemented a Leaf system.

USA TODAY currently uses the Leaf system to edit the images that come off of the Associated Press (AP) wire service—from cropping to sizing—then loads them into the Scitex system for placement on the page. Rob Aronson, Director of Imaging for *USA TODAY*, explains what having the Leaf system has done to bring *USA TODAY* more into digital production:

We have hard copy scans, we have scanners interfacing to the Scitex system. We've got the graphics coming into the system. We've got the wire photos coming into the system. That is what really made it successful for us to really begin to bring the newspaper into an elec-

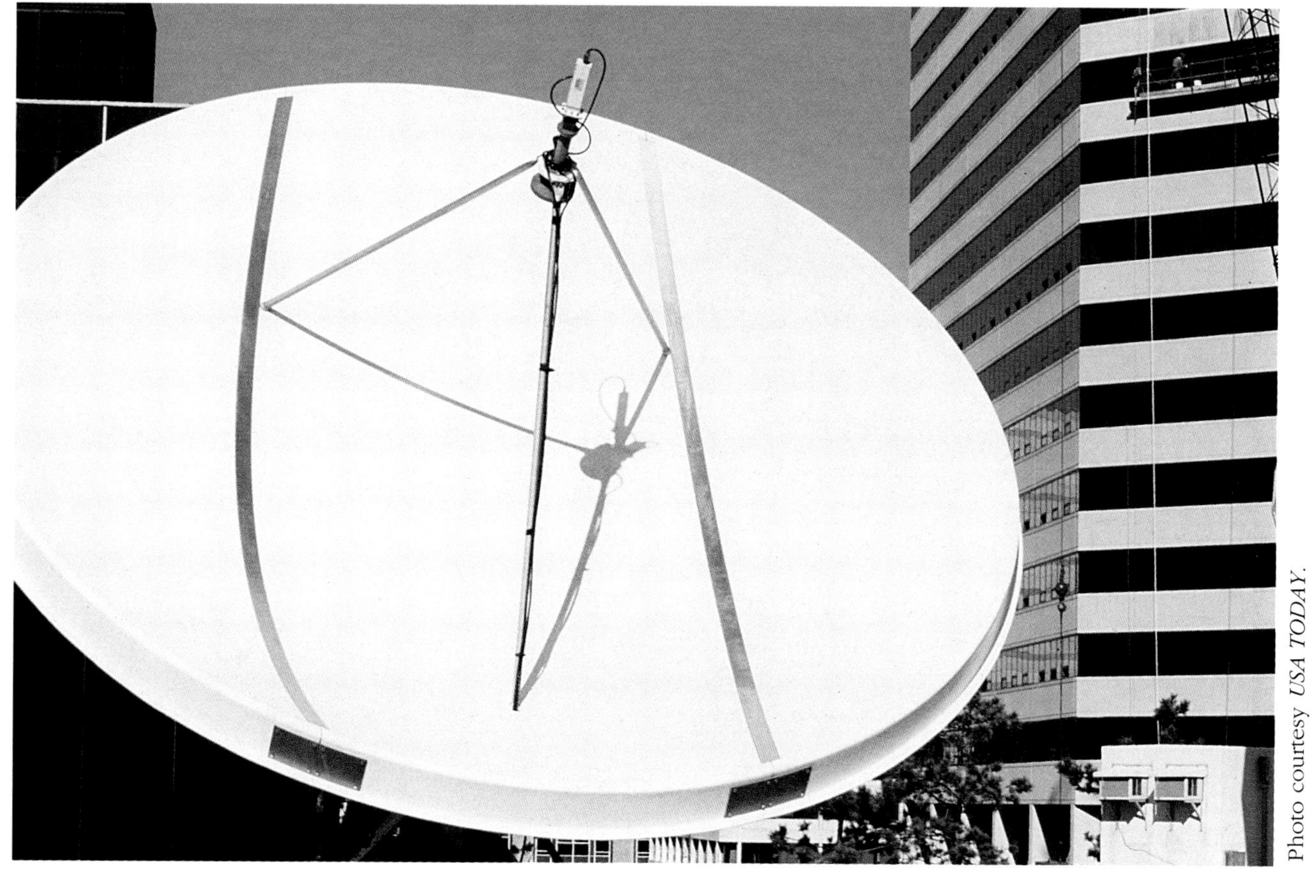

Photo courtesy *USA TODAY.*

USA TODAY *uses a Ricoh facsimile system to simultaneously transmit data via satellite to 32 remote print sites.*

tronic form. Now we have a long way to go. We're not near where we'd like to be.

Merging Text and Graphics

The black and white photo output usually originates from a wire system, then is output from the Leaf system as halftone prints at a line screen of 85 dots per inch (dpi). Those halftone prints are pasted directly onto a page, and the page is shot as a composite.

Color information is output from the Scitex as negatives. Once these separations are complete, the black portions for color pages are merged with the text portion by a double-burning process, then converted to Veloxes, which are essentially transmission prints, for distributing to their 32 remote printing sites.

Their method for transmitting to their remote print sites is a Ricoh facsimile system that they have had since 1982. They are entirely self-maintained with it, and are even at the point where they are building their own parts and have their own engineers.

This facsimile system simultaneously transmits a page to all 32 sites in what is called broad-

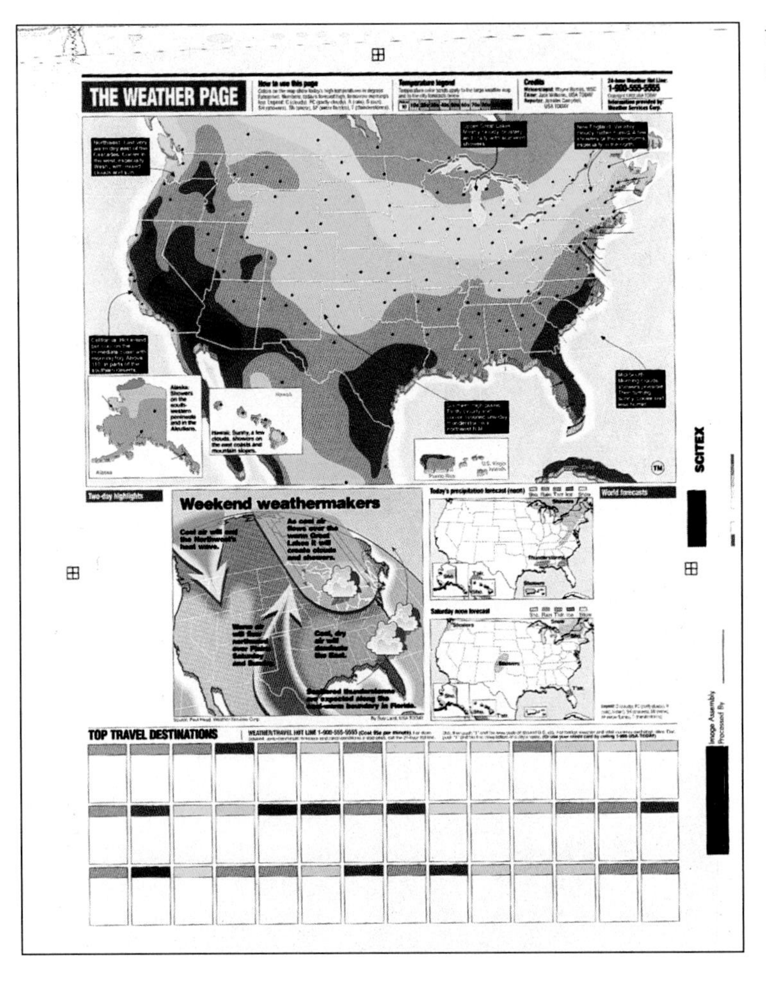

USA TODAY *uses color keys, like this one for the weather page, to proof for accurate color.*

cast mode. The average transmission time for a page is four to six minutes. The system works just like a very high-resolution fax machine that you would find in an office, except transmission is via satellite, not phone line. At the other end, instead of a paper print coming out, they'll get a 13" x 21" negative coming out with registration marks. That negative is used to burn the plate.

Color Proofing

The Editorial department has dye-sublimation and thermal-wax color printers for color proofing, available for anyone who wants to look at something early on. However, finished color proofs are not readily seen and available for everyone to look at until it is actually almost ready to transfer. At that time, a color key is produced. Although a color key is not always an accurate representation of how an image translates into print, a skilled observer knows what to look for and determine how an image will actually translate to the final product.

For advertising, everything goes to a four-color proofing press on the sixth floor of the *USA TODAY* building. This way, *USA TODAY* can show

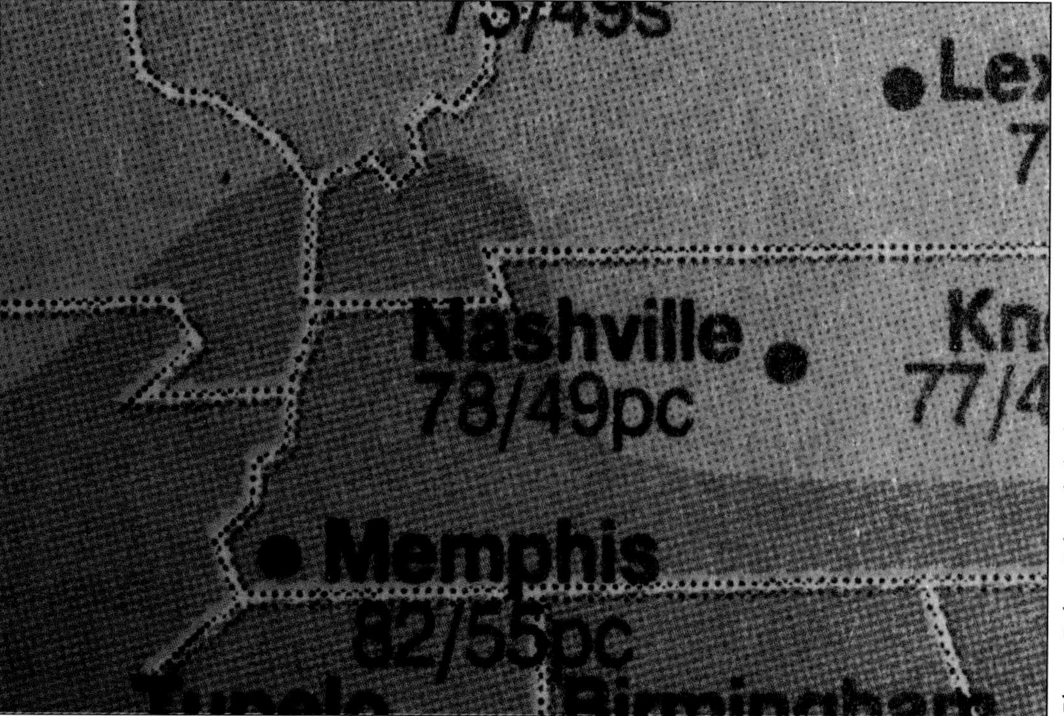

Photo: Robert Edward Sherbow

USA TODAY *is printed at a line screen of 85 dots per inch, which is the coarsest screen that can render a clear, crisp image.*

an advertiser what their ad is going to look like in the newspaper, because there is really no off-press proofing system for newsprint.

Another reason for generating advertising proofs off of a printing press is to send them to *USA TODAY*'s remote print sites. The proof press prints about 300 of each ad going into the paper, then distributes a handful of each to each site. When a print site runs the ads on press, the pressmen have some sort of point of reference. They know what the advertisers expectations are so they can have some sort of consistency. There are a lot of other quality control steps, but that's the final check when the actual piece is printed. They have to have a reference as to what the final product should really look like.

Line Screen

USA TODAY produces near-magazine quality color photographs and graphics. A line screen of 85 was chosen because it is probably the coarsest that renders an acceptable image while holding to highly absorbent and porous newsprint.

Another reason involves transmitting data over their facsimile network. To release a news-

THE NATION'S NEWSPAPER

INTERNATIONAL

USA TODAY

Money

TUESDAY, OCTOBER 20, 1992

MONEYLINE

TODAY'S TIP-OFF

Chrysler's return

Fixed mortgage

THE STOCK

5 YEARS AFTER THE CRASH

Stocks: A str

COVER STORY

New r shoul crash

VIA SATELLITE

THE SHAQ: MICHAEL'S CO-STAR

No. 4 KANSAS UPSET BY No. 23 OKLAHOMA 80-77

USA TODAY

NO. 1 IN THE USA . . . FIRST IN DAILY READERS

50 CENTS

EASTWOOD, PACINO HEAVY HITTERS IN OSCAR RACE

THURSDAY

THE PLAN AT A GLANCE

TAXES

GOAL: Bring in $328 billion over five years

STIMULUS

GOAL: Boost the recovery, invest long term

CUTS

GOAL: Trim deficit by $704 billion in five years

THURSDAY, FEBRUARY 18, 1993

NEWSLINE

'Focus on the future'

Clinton calls on nation to 'unite and act'

POLL: 79% support plan

USA TODAY *has been printing with environmentally responsible soy-based inks for over four years. These inks adequately produce* USA TODAY*'s vivid colors.*

Photo courtesy *USA TODAY.*

paper on time, the facsimile has to scan at a certain sampling rate. If *USA TODAY* were to produce its photographs and graphics at a higher screen ruling, they would have to raise the sampling rate, which would slow down the scanning time of the facsimile and create a large bottleneck at the distribution phase.

Images with a relatively coarse line screen are quite acceptable, since *USA TODAY* is a newspaper. In fact, anything greater probably would go unnoticed by the reader.

Environmental Considerations

For the last four years, *USA TODAY* has been gradually transferring from oil-based to soy-based inks, which adequately produces *USA TODAY*'s vivid colors. They also use as much recycled newsprint as possible, however, it is hard to find. There is also no standard that indicates how much recycled material is added to paper labeled "recycled," or whether it is post-consumer waste or pulp from the floor of a paper plant.

USA TODAY currently uses about 40% recycled paper and is purchasing as much recycled newsprint as it can. In fact, *USA TODAY* is rewarding companies that manufacture recycled newsprint by giving them larger contracts.

Photo: Robert Edward Sherbow

EQUIPMENT

- Scitex, Crosfield, Hyphen, Iris, and Leaf graphics and prepress workstations and peripherals (high-end and direct PostScript devices)
- Macintosh workstations
- Assorted hard drives
- Assorted backup drives
- Assorted scanners
- Dye sublimation and thermal wax transfer color proofing printers
- Four-color proofing press for advertising
- Proprietary software for the high-end systems
- QuarkXPress, Aldus PageMaker, Adobe Photoshop, Color Studio, FreeHand, and Adobe Illustrator for the Macintosh systems

CONVERSATIONS

Photo: Robert Edward Sherbow

ROB ARONSON
Director of Imaging

Rob has been doing imaging work since 1984. He studied at RIT, and did installations and training for Crosfield before joining* USA TODAY. *While at RIT he gained a lot of experience with the Scitex system, as well as with quite a number of scanners. He joined* USA TODAY *about four years ago, just as the company was bringing their Scitex system on line.

There are still a lot of parts of the operation that are still manual, and there's a reason for that. We had a rather large Scitex system here that really wasn't being used for the newspaper at all. All the scans and all the color was being randomly output. All the graphics for *USA TODAY* were created manually by artists with many many overlays, sometimes as many as 40 overlays, to create a graphic effect. That took a lot of personnel to put that together because *USA TODAY* is very graphically oriented.

Probably the main thing that made us successful in going electronic was the fact that we were able to take the color graphics onto a Macintosh platform and instead of having an artist create a very complex manual type graphic, they can put all their energies and creative efforts into creating something on a Macintosh. It's about that time that Scitex came up with the Gateway and the Visionary Interpreter for PostScript, of which we were a big proponent. When these systems became available we could import the graphics directly into the Scitex system.

So we really wanted to have a PostScript interpreter for the Scitex system so we could pull the graphic elements into the system. At that time also we were experimenting with some electronic wire photo systems which we've seen grow tremendously in the last few years.

There are actually a number of Mac networks that have developed over the years here. We're trying to unify. There's actually a big project going on right now.

A lot of people are saying "Well, in order to be really truly efficient, you need to have a fully electronic paginated product." To me, when you do something like that, you are making the technology the end instead of the means. Technology always has to serve the problems. There always has to be an advantage to do something. If I say I want everything to be electronic, there has to be a business reason for making it so.

SUMMARY

When *USA TODAY* was launched in 1982, many people were skeptical that a national newspaper with a four-color layout, more like a magazine than a traditional newspaper, would be successful. People were not sure a newspaper of that nature could compete with traditional regional publications. Now, a simple glance at many metropolitan newspapers—which now include color and sophisticated computer-generated graphics—indicates that it was an idea worth copying. The staff at *USA TODAY* have both transformed the industry and led the way. They expanded the scope of a newspaper to techniques being used in other industries, compiled them in an efficient, coherent manner, and created a national product.

CHAPTER 4

BOOK PRODUCTION

This chapter discusses the creation of a four-color book. Learn how MIS:Press and the authors used desktop technology, such as style sheets and master pages, to streamline the production process.

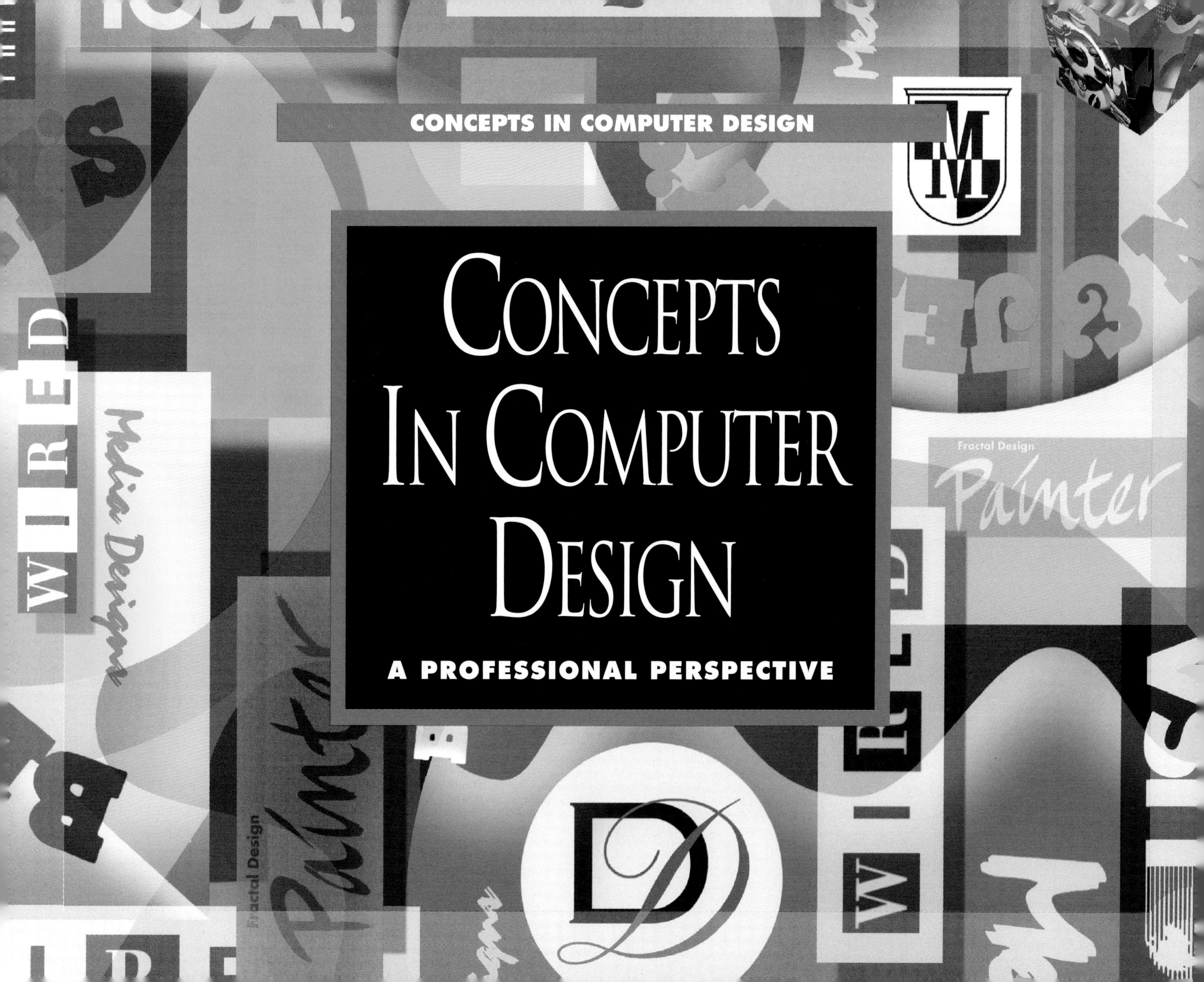
CONCEPTS IN COMPUTER DESIGN
CONCEPTS
IN COMPUTER
DESIGN
A PROFESSIONAL PERSPECTIVE
WIRED
Media Design
Fractal Design
Painter
Fractal Design
Painter

As the first four-color book published by MIS:Press, numerous meetings were held to ensure production and printing would run smoothly.

INTRODUCTION

Dawn Erdos and Leslie Singer developed the idea for this book and approached Cary Sullivan, Senior Development Editor at MIS:Press. Cary got very excited about the project, and brought it to Publisher Steve Berkowitz. Several meetings followed with printers, production staff, and sales staff to further refine the concept of the book and determine what was needed from a technical point of view. Cary explains:

Since this was our first four-color book we wanted to make sure that everything was going to be in place from a printing perspective. We had a representative from the printers—Arcata Graphics—come in and talk to us about what we needed to do, such as get a template together for them to test, that sort of thing, to make sure that everything would run smoothly on the color side.

The book was unusual for MIS:Press for two reasons: Its subject matter was broader than most of its books, which usually are more technical and tutorial, and it is very unusual for the authors, rather than the publisher, to be handling the design and production of the book.

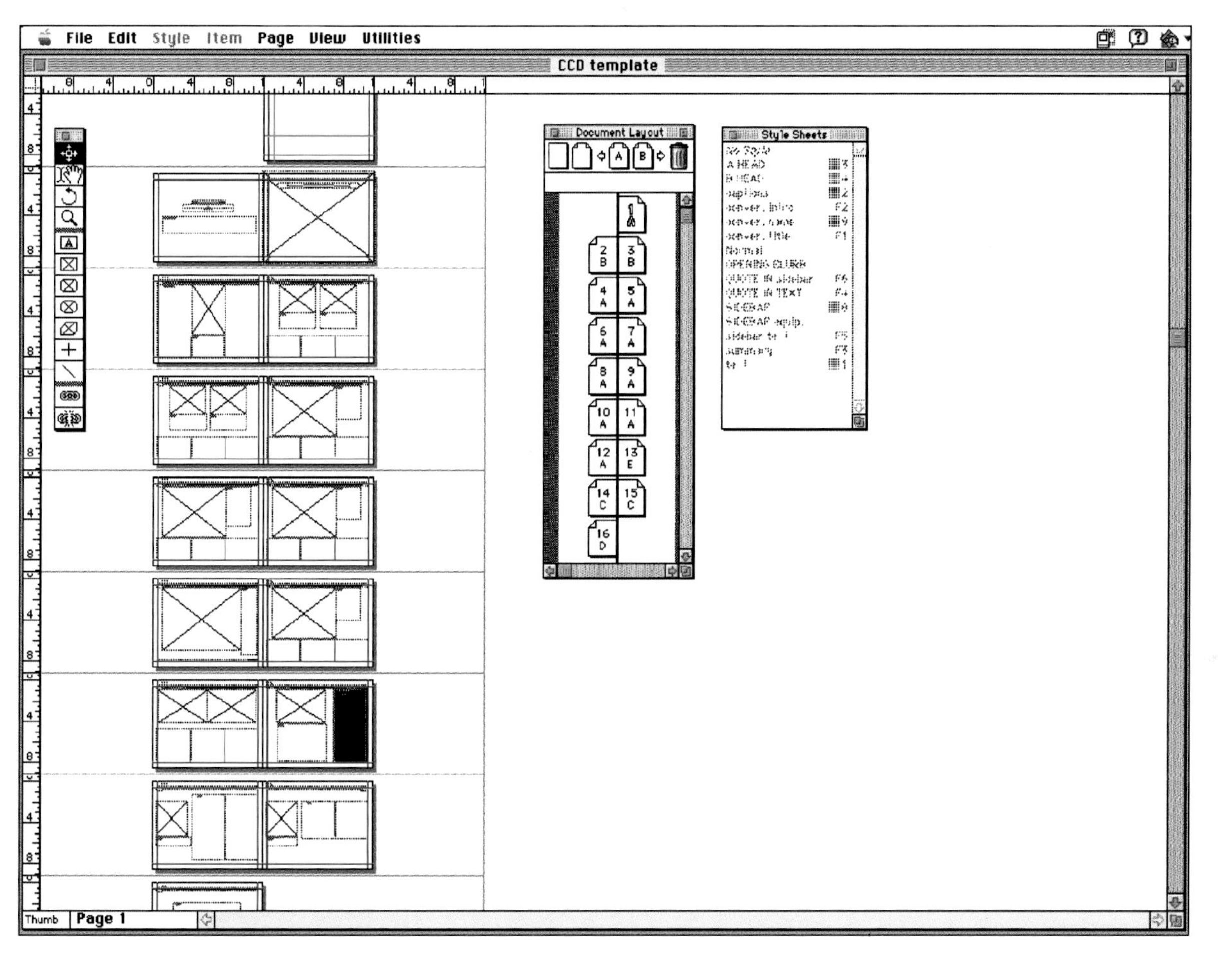

Art Director Diane Cronin created a template in QuarkXPress based on the sample chapter. Her use of style sheets and master pages significantly streamlined the production process.

ELEMENTS OF THE PROCESS

The first step in producing the book was to compile information about the industry, then break it down into manageable categories and contact people to see if they were interested in participating. After that came a series of about 50 interviews and writing of the chapters.

While the chapters were being written, the designers and companies covered in the book had to be contacted to collect enough visual material to create a good balance between editorial and visual interest.

As artwork was being collected, Dawn and Leslie began to create a format that worked for everyone from the editor, to the sales force, to the printer.

As the first four-color book produced by the publisher, the authors and staff at MIS:Press had a conference with the printer to balance out the issues of cost, specifications, schedule, and electronic formats.

After the conference, Leslie and Dawn worked out the style of the book and created a sample chapter for a print test. At that point, Diane Cronin was brought into the project as an art director. She set up a template and style sheets in QuarkXPress and began coordinating the production of the book. Leslie explains

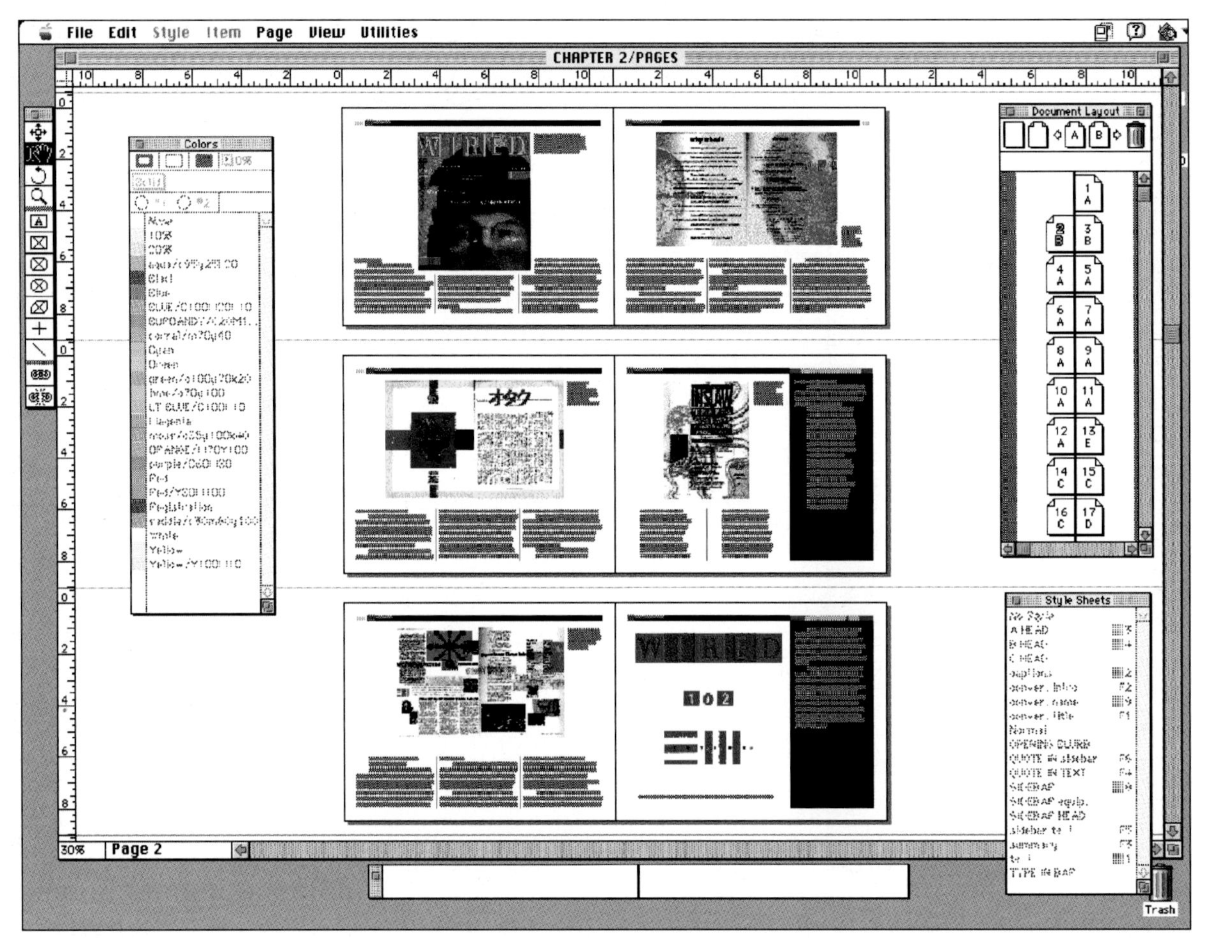

Once the template had been created, Diane placed the graphic elements and flowed in the text.

For the publisher's needs, we also had to talk to the printers, and the printer and the publisher had to look at costs, then get back to us to make sure that we weren't going to have any bleed pages except for the intro to each chapter.

From there we set up a style sheet and a grid to insure consistency throughout the book. The production end was very simple, and using a program like QuarkXPress, there's no reason for it not to be.

Basically Dawn and I created the look and feel and style of the book, and Diane is really pulling together all of the elements. She created the style sheets and master pages, and did most of the hands-on work to pull the book together.

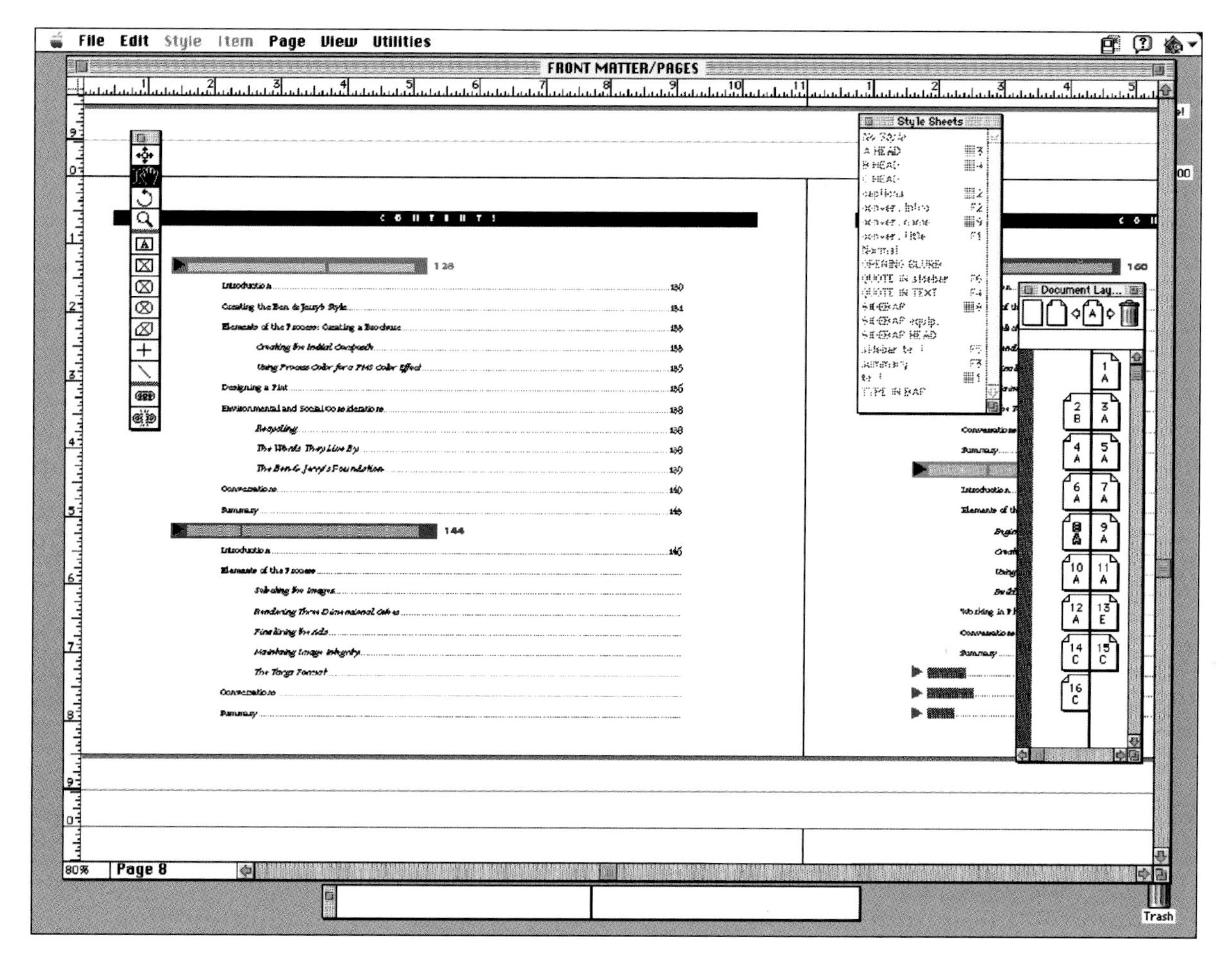

One of the last steps in production was to paginate the book and format the Contents.

Once the design of the sample chapter was complete, Diane flowed the text from the edited chapters into the QuarkXPress template.

Next came placement of the graphic elements. The graphics had arrived in all different forms—from electronic encapsulated PostScript (EPS) or TIFF files to veloxes to C prints and chromalins. Diane had to catalog and organize the visual material before she could incorporate it into the layouts. The electronic files simply were placed into the layouts, while the non-electronic materials were scanned in color at 75 dpi to 300 dpi for position only (FPO), then placed. These images would later be scanned on a high-resolution drum scanner at the print house and be replaced in the QuarkXPress files just before the files were ripped.

The layouts were sent to Cary for editorial review. The corrections that came back were incorporated into the QuarkXPress files, and the entire book was paginated.

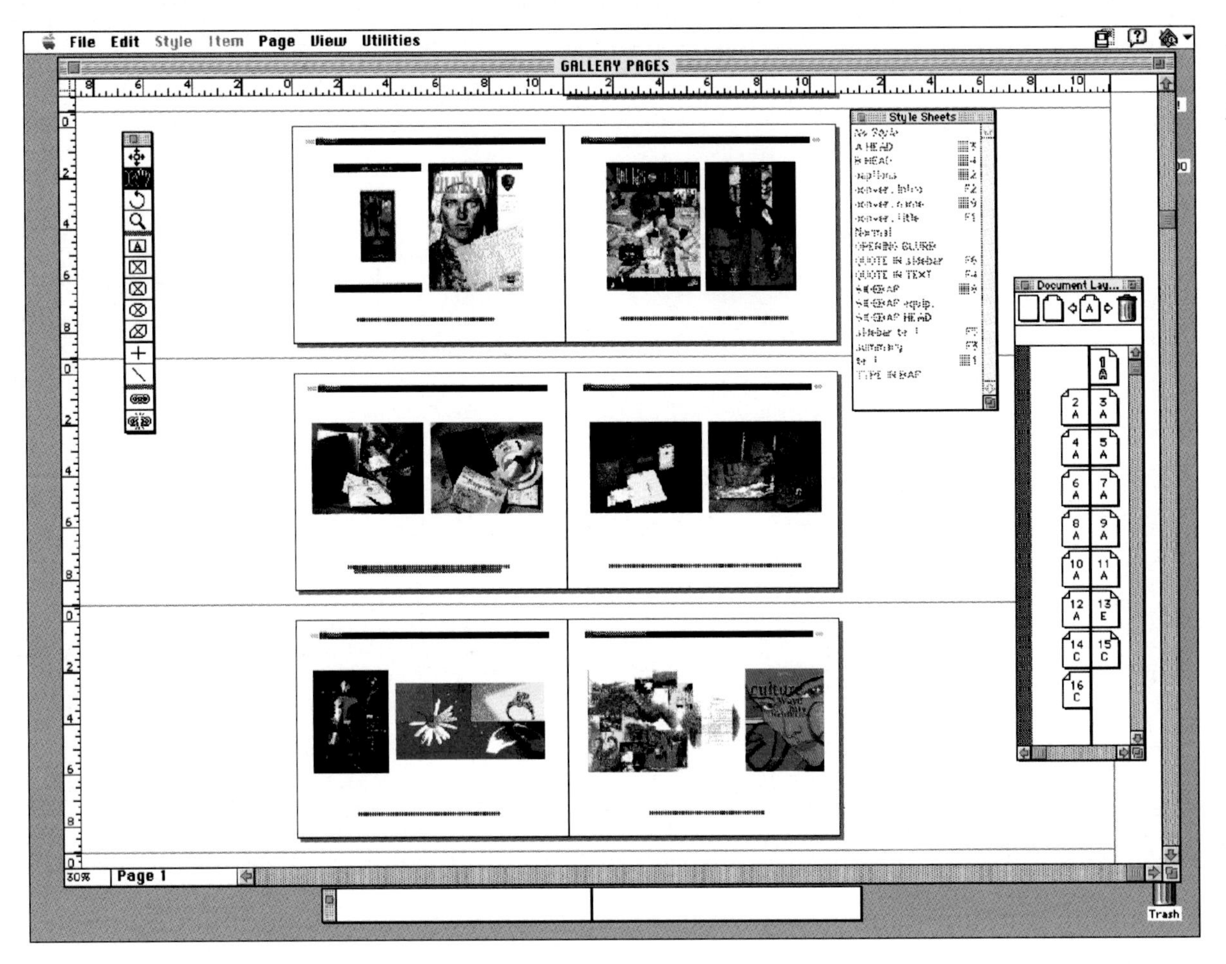

More than 100 pieces of art were scanned for position (FPO) and cataloged before the book could be released to the printer.

After finalization of the book, Diane copied the files onto 44Mb SyQuest cartridges and sent the cartridges, along with the loose artwork to Kurt Andrews, Production Manager at MIS:Press.

As the printer liaison, Kurt further organized and cataloged the artwork and electronic files, and forwarded them to the printer. The printer scanned in the loose art, replaced the FPO art, and ripped the entire book through a Scitex high-end prepress system.

For approval, blueprints were made of all pages, and the color material was output as chromalins. The book went to press on a four-color offset press on glossy paper stock in 16-page signatures.

Working with the Printer

One of the biggest production issues faced by MIS:Press was finding a printer that has enough technological competence to handle a four-color, electronically imposed book. Kurt Andrews, Production Manager for MIS:Press, explains that many printers are hesitant to make the investment required, because they are already invested in the

Photo: Daniel Williams

The artwork arrived in all shapes and sizes—from chromalins, slides, and veloxes to EPS and TIFF files.

older technologies, and the current technologies may be phased out in the next few years:

> *Some printers are really afraid to make that investment because they see the future of printing within the next year or two basically just going from disk to plates, and eliminating the whole film process. So they're dragging their heels as far as investing in an imagesetter and similar equipment. When the technology is there to go just to plates, then they'll make that investment. I think that's wrong. If they don't understand imagesetting technol-*

WORKING FROM REMOTE SITES

The New York offices of Singer Design was the home base for the production of the book. Both Dawn and Leslie, however, had a need to work from remote sites as well. Dawn is based in Virginia, and Leslie had a business commitment in Europe during part of the production process.

Leslie hired technical consultant David Acosta to link everyone together. David selected high-speed Global Village modems for data transfer. Using fax machines, modems, and telephones, Diane, Leslie, and Dawn were able to work on the same files, without having to be at the same location.

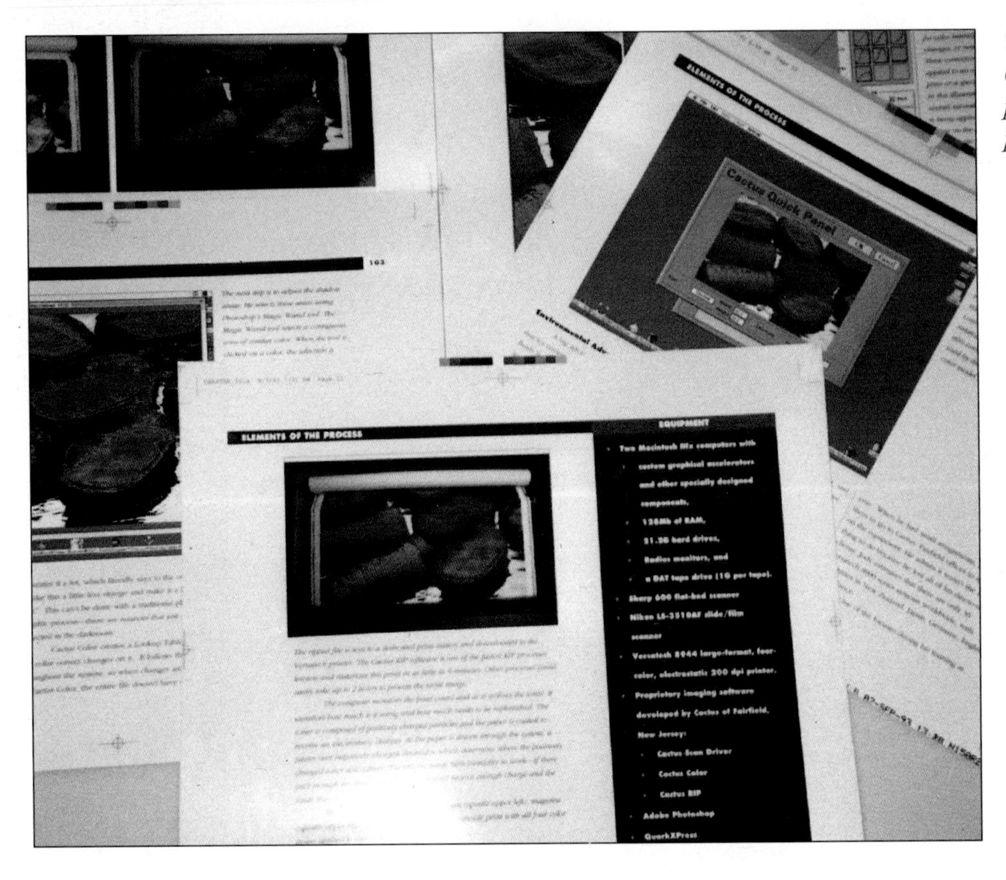

The printer generated blueprints ("blues") for proofing text and position (left), and chromalins for proofing color.

ogy, if they haven't gone through the stage in between where we're going from disk to film, when the technology starts going directly to plates, I think it's going to be a really foreign concept to them and we're going to have a lot of problems.

All MIS:Press books are delivered to the printer on disk, but most of them are one- or two-color books, sometimes with one or two signatures in full color. This was the first book that used four colors throughout published by MIS:Press. For that reason, Kurt says that it was critical for the printer handling this job to be on the cutting edge of technology. He selected Arcata Graphics for that reason.

Although he tries to work on the cutting edge of this technology, Kurt says that it is necessary to keep learning, and he hopes print houses do the same.

Photo: Daniel Williams

EQUIPMENT

- Five Quadra 800 Macintosh family of computers
- Macintosh IIci network server
- Macintosh IIsi remote-site workstation
- Apple Powerbook 170C notebook computer with a color monitor
- Five 19" RasterOps 256-color monitors
- Ethernet network
- Tektronics PhaserJet III thermal wax transfer color printer
- Apple LaserWriter laser printer
- UC630 UMAX 600 dpi color scanner
- Five 44Mb SyQuest removable media drives
- Global Village Teleport Gold 14.4 Kbps external fax/modems
- APS DAT tape backup system
- Adobe Photoshop
- QuarkXPress
- Adobe Illustrator
- Aldus FreeHand
- Fractal Design Painter

CONVERSATIONS

Photo: Daniel Williams

LESLIE SINGER
Author and President, Singer Design

Leslie studied design at Parson's School of Design (New York). She became an advertising art director and worked at a number of New York advertising agencies, including the Marschalk Co., Ogilvy & Mather, and McCaffrey and McCall. She loved print work, but discovered she couldn't do the kind of work she liked in a corporate environment. In 1990, she purchased a Macintosh and founded Singer Design.

I think that if you don't have this technology, you're not in the business anymore, and if you are you're hanging on by a thread. As far as my business goes, it's made me love my craft more because it's kept out some of the grueling portions of it, like magic marker comps and having to rely on typesetters and outside services like stat houses. It has allowed me to have more control over the creative process right through production. That was something that an art director was not part of in the past. I thought it was really a nightmare. I loved my craft, but I didn't like the business of doing it, and now with the computers it's turned around 180 degrees. It's so streamlined that it's made my business possible and it's made me a nicer person.

I would never have considered going into business for myself at the time I did if it weren't for the technology. Basically, all those outside services ate into your profit margins, and there was too high a cost in revisions. For example, if the type shop gave you a widow at the end of your paragraph, and you sent the galleys back, they charged you for it. If you didn't like the tracking or the kerning, you used to have to sit there with a razor blade and move each letter on a board. If the stat house cut off part of your shot, they may not charge, but it cost you time. Those were all issues that I found overwhelming. It was hard enough to take care of clients and do good work, plus manage a crew of suppliers. I have always felt that the more that you have to rely on other people, the more trouble you have.

I think it's very exciting the way the industry is changing and becoming electronic. I think that we have to stand together as art directors and not become production managers—I feel very strongly about that.

Photo: Darshan K. Khalsa

DAWN ERDOS
Author and President,
Solstice Communications, Inc.

Dawn has worked as a writer, editor, and graphic artist for ten years. In 1990, she founded Solstice Communications, which provides editorial, production, and packaging services to the publishing industry. She began working with desktop technology in the mid 1980s, and has been sitting at her computer ever since.

If it weren't for this technology, I wouldn't be able to have my own business. I would probably be working on staff at a publishing house as an editor or production manager. The computer has let me work hundreds of miles away from my clients, and to actually participate in the production of this book hundreds of miles away from where the bulk of the work is taking place. Plus my overhead is much less now.

I think that there still is a bit of a prestige element of having your office in New York, or Los Angeles, or even San Francisco. If you're working and living out in the country in Virginia, you do lose a little bit of that status element, but once more and more people discover—and they have been—that you can basically work from anywhere there is electricity, a phone line, and Federal Express pick-up available, that situation will be changing and people will be living in New York because they *want* to live there, not because they have to live there. If they want to live and work elsewhere, they are able to.

It's faster for me to get a file now from Leslie's office over the modem than when I was living across town in Manhattan. It takes less time for me to get a file over the phone line than it did for me to get in a taxi and go to her office and pick one up, or even to use a messenger service.

CARY SULLIVAN
Senior Development Editor, MIS:Press

Cary began learning personal computers in the early 1980s while she was working at a Japanese trading company. From there she moved to John Wiley & Sons where she started as an Editorial Assistant in its Encyclopedias department. She happened to work on some of the projects with the head of Wiley's Digital Production department, which at the time (1987) was very new. The technology fascinated her, and she applied for the next opening in that department. She stayed there until 1991, when she took a position with MIS:Press as a Development Editor.

If anybody had ever told me while I was still in school that I would eventually be working with computers I would have thought they were nuts. I never had a class in it, I just happened to fall into it. I got some computer training in spreadsheets and word processing at a Japanese trading company that I worked for. That opened the door to a job at John Wiley & Sons, which in turn opened the door to desktop publishing. I just love the technology. It's amazing to see how quickly it's grown, and how sophisticated it's gotten in the last few years. It has made a tremendous difference in my career and it opens up the doors for so many more people and possibilities. It lets you do things faster and better and cheaper than ever before.

A perfect example of how computers have changed the industry is what's happening at print houses now. Just a few years ago there were very few printers who could do electronic imposition, but because the customers have created a demand for the electronic imposition (since it's faster and theoretically cheaper than the traditional methods), the printers have had to accommodate clients by using the technology. Just the fact that we can typeset with computers rather than using traditional methods has created six jobs at MIS:Press. The changes in desktop publishing have enabled us to do things in-house that we never could have done before, and have added a dimension of creativity that wouldn't have been possible otherwise.

Photo: Daniel Williams

DIANE CRONIN
Art Director

Diane studied art and graphic design at Radford University, Virginia. She is currently an art director and designer for Singer Design, working on projects ranging from annual reports and corporate presentations to magazines for children.

I never had to work manually. When I graduated from college, the industry was already on computer. I doubt I would be an art director if it were otherwise.

When I came into this project we only had a sample chapter that didn't have real art or real copy. I took that chapter and used it as a guide for where the page numbers are placed, what the margins are, use of graphic elements in every chapter. I took the sample and made a basic template. From there I tailored the original layout to each specific chapter. I developed each chapter individually, because each chapter has it's own individual story, with different art, different requirements, and different layouts for every page.

The template I created had a basic opening spread for a chapter, and some elements of the center pages were similar. Then a few pages at the end were "conversation" pages. I simply made a master page for each different one, revised them as I found necessary through working with it, then created style sheets. The hardest part of the project was setting up the template and the style sheet. Once I got that down, the rest of the project was much easier.

KURT ANDREWS
Production Manager

Before I came to MIS:Press, I worked at Van Nostrand Reinholt as a Production Manager. I worked with their architecture and graphic design books, so I worked a lot with designers and architects. I worked with their original art as far as getting that separated and as far as their designs.

A lot of the designers are graphic artists who are really into the desktop—graphic design on the desktop and electronic publishing. So I learned electronic publishing through the authors who worked at Van Nostrand.They were producing these files and they'd be saying, "I created this as a TIFF and I did this piece of art in FreeHand and I'm going to supply it to you on disk." I'd never heard of these things. I'd never heard of anyone mention a TIFF or FreeHand or EPS. The first time I got on the phone with a designer and he started babbling away about EPS files and TIFF files, it was almost like a foreign language. So I had to learn it.

Sometimes I still feel like I'm not even up to speed with it, I feel like I've got the jargon down—half the battle is understanding what the jargon means—but I feel like it changes so much and from day to day and there's always something new that every time that I get caught up, there's a new program or something new out there for me to learn. I don't feel like you can ever really be caught up on it.

I'll always feel like there's something more to learn out there. It's challenging, and if your job isn't challenging, what's the point?

SUMMARY

The recent strides in desktop publishing, printing, and communications technology are what made this book possible—not only in terms of concept, but in production methods as well. The technology opened up design options in a way that enhanced creativity and productivity, which is, after all, what this technology is all about.

CHAPTER 5

CORPORATE IDENTITY

This chapter focuses on the creation of an identity for Medallion Hotels. Learn how Diana DeLucia Design shaped this hotel's identity, from its name to its printed materials.

DIANA DeLUCIA DESIGN/MEDALLION HOTELS

MEDALLION

HOTELS

Diana DeLucia Design began working with the hotel's parent company two years before the first hotel was developed. The firm began creating an identity for the first Medallion Hotel six months before its grand opening.

INTRODUCTION

The focus of a corporate identity campaign is to present a company's individual personality. Often, this begins with the creation of a logo or identifying mark, but can easily extend to any material that is generated by, or representative of, the organization, such as all printed matter, video presentations, pens, clothing and uniforms, and much more.

Some of the most evident and extensive corporate identity campaigns are found in hotels. Any respectable hotel displays its identity on everything from its advertising, to signage, room keys, menus, robes, phone directories, and soap.

If possible, a hotel develops its identifying materials before it even opens its doors. When the client founded the Medallion Hotel group, he contacted Diana DeLucia Design, whom he had met when she developed the identity campaign for Rock Resorts' glamorous Hanbury Manor in England.

LEGACY	**LEGACY**	LEGACY	LEGACY
MEDALLION	**MEDALLION**	MEDALLION	MEDALLION
COMMONWEALTH	**COMMONWEALTH**	COMMONWEALTH	COMMONWEALTH
WATERMARK	**WATERMARK**	WATERMARK	WATERMARK
LANDFALL	**LANDFALL**	LANDFALL	LANDFALL

Diana DeLucia Design began the identity campaign even before a name was selected. Part of the naming process involved a type exploration to see how each option looked in print.

ELEMENTS OF THE PROCESS

When DeLucia accepted the project, the client had just founded a hotel-management company and had not yet begun doing business. The group posessed no existing hotels.

Naming the Hotel

Part of the project included the development of a name for the company's first hotel. Rather than the luxurious image of Hanbury Manor, the client wanted an upscale, professional businessperson's image. With rooms starting at only $125 per night, the client wanted a value-oriented, more masculine look, without appearing overly frugal or ordinary.

Diana DeLucia, president of DeLucia Design, explains that the first hotel acquired was much more upscale than she had expected. She explains:

I think one interesting thing with this was when we developed the name, we never knew if there would be a hotel. We just knew it

Diana DeLucia Design began its exploration of marks with the thought of using a crest or coin motif.

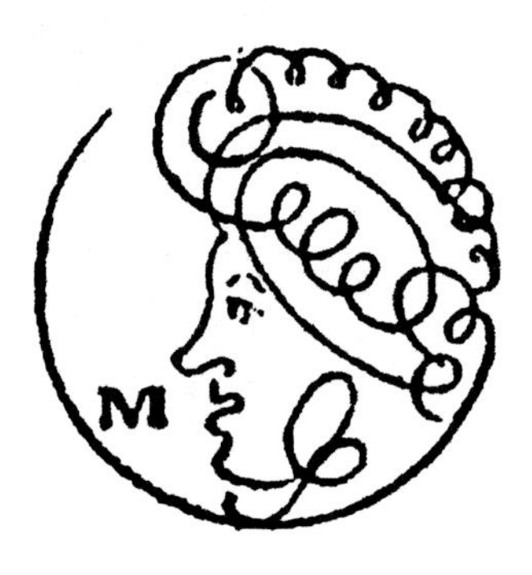

was a management company for hotels and the colors were blue and gold. That's what it was because it was a corporation. It was just stationery with potential for something or other. And then when we finally got one real hotel and I went there, I said "We can't use blue and gold!"

When I went there and saw the beginnings of renovations, it's pretty hip kind of furniture, muslin covered sofas. It's not really square. So we thought we could push it a little more than blue and gold on white paper.

That's why her firm chose recycled, textured papers with an uncoated look and copper ink.

Once the hotel in Houston, Texas, was acquired, and the basic image began to be developed, they had to find the right name to fit the image. Diana describes the process:

We sat around and talked about names. He [the client] actually came up with the name Medallion. We worked on a list of names for him and showed him

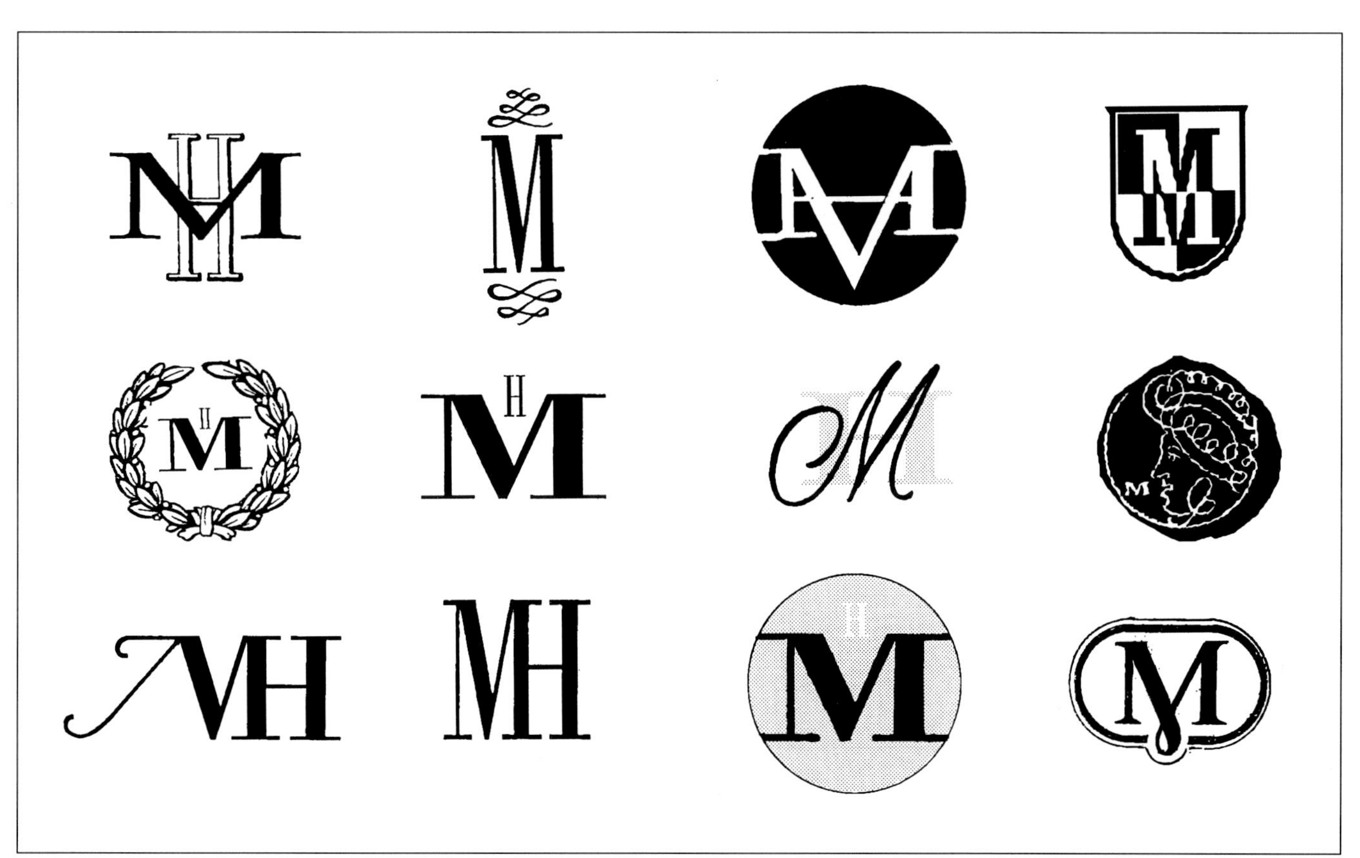

The crest and coin concepts were further refined and modernized.

which names would work and recommended two of them and then did programs for both. Sometimes with a name you have to see it and feel it .... and then we ended up with Medallion. Since then they've expanded and they now have six or seven properties. So we've been doing a lot these. From this one we applied it to another hotel that they own and we've had to modify it and do it in different colors. So it is an ever-growing sort of thing. We probably have about 100 files in the computer on it. It's one of those things that if you had to do it by hand, it would be ridiculous.

Next, they had to determine the specifics of the name. The client had simply said to them to use "Medallion Hotel." They had to figure whether to use "Houston Medallion Hotel," "Medallion Houston Hotel," or "The Hotel." Or, if they were just to use "Medallion," would people know it was a hotel. While that was happening, they had to begin thinking about what the materials would actually look like.

MEDALLION	MEDALLION	MEDALLION	MEDALLION
MEDALLION	Medallion	medallion	MEDALLION
MEDALLION	MEDALLION	Medallion	MEDAL-
MEDALLION	MEDALLION	Medalli	Medallio
MEDALLION	MEDALLION	MEDALLION	MEDALLION
Medallion	MEDALLION	MEDALLION	MEDALLION
MEDALLION	MEDALLION	MEDALLION	MEDALLION
MEDAL-	MEDALLION	MEDALLION	MEDALLION

Once the name was selected, the type exploration was begun.

The Exploration

Diana DeLucia Design performed an extensive exploration for the logo. One of the most important considerations during the exploration was the possibility that the logo would appear in various sizes, on various media, so it had to be flexible. In fact, the logo was eventually printed on items as small as a bar of soap, and as large as a 59" sign.

Setting the Type

The first step was to begin setting type, to find a look that worked. They used a large variety of typefaces, using uppercase letters, lowercase letters, specially refined character sets, and more. Barbara Tanis, the designer in charge of the project, explains why they chose Bodoni and Bank Gothic:

> *The Bodoni was a classic typeface with beautiful letter form for that particular word. We condensed it partially to make it proprietary and also because it's such a long word and we didn't want to have all the letters touching.*
>
> *They seemed to look more elegant and distinctive when they were condensed and then spread out.*

MEDALLION
HOTELS

HOUSTON
MEDALLION

After the typefaces and mark were selected, Diana DeLucia Design went through an additional exploration to combine them for the final logo.

> *Bank Gothic was a sophisticated typeface and it appealed to the sort of executive market we were trying to reach. And it wasn't as formal as doing everything in Bodoni. It was more of the feel of the visuals of the hotels. It was contemporary, but not too contemporary.*

Patricia Kovic, one of the design directors on the project, describes how the two chosen faces work together:

> *There was a visual play. One of the useful things about it was that sometimes with a logo like the word Medallion, that is the logo. If you start doing every one in Bodoni you start diluting the effect of the proprietary Bodoni, so we wanted a different typeface and one that could stand alone. That type of treatment. But then, the third type-face that we used is regular Bodoni, a little condensed.*

Creating the Mark

Next, they did a lengthy exploration on marks (the image that goes with a logo). DeLucia

Diana DeLucia Design created several treatments of the mark to give a variety of options for collateral goods.

Design created a wide range of marks and worked with them for quite some time, until the personality of both the corporation and the hotel were more firmly in place. Some of the marks were generated on the computer, and some of them were hand-drawn. Barbara explains that the computer is merely a tool to them, and as a tool is not always appropriate for certain situations:

> *We have a view here, although some people don't agree, that everything must not be done on computer. That these are very important vehicles and we use them a lot and sometimes you can't communicate what you want on the computer. So we frequently use sketches, combine them with the computer, scan them in, do a lot of thinking with hand work.*

They finally presented two marks with which they were comfortable to the client. Once the client made his selection, they refined the mark into its final form. Its first appearance was on the 59" sign placed on the outside of the hotel, so there was a redrawing of the logo in varying sizes on the computer to make sure it worked.

Varying the Treatment

Since the logo would be displayed on so many different items, they created two additional mark treatments based on the primary one to create variety and avoid a stagnant system. This was also done to be sure the mark used on a particular item was appropriate to the subject while maintaining continuity.

Completing the Campaign

DeLucia Design had only three months to complete the identity campaign because the client had set an opening date for the hotel that was not going to change. In that three month period, DeLucia Design had to develop the logo, soaps,

AN IDENTITY FOR THE ENTIRE GROUP

Since the acquisition of the first Medallion Hotel, the chain has acquired additional properties. In three cases, the name "Medallion" was implemented immediately, and DeLucia Design simply had to change the name of the cities on the collateral material.

In one instance, the group acquired a fully operational hotel and decided it would be most beneficial to gradually change over to the new identity, to avoid confusion and alienation of existing clients. The name of the hotel was kept initially, and DeLucia Design adapted the Medallion style to the existing image. Their job is to, slowly but surely, marry the images of the hotel and its new chain without ever using the name "Medallion." Once the Medallion image is fully associated with the hotel's image, they can begin introducing and promoting the "Medallion" name.

Room Number Accomodations Arrival

Account No. Status Guests Departure

Room Rate

Guest Name

Company

Address ☐ Business ☐ Residence Telephone

Address

City State Zip Code

Package Plan

Group Affiliation

Your Travel Agent Advance Deposit

I agree that my liability for this bill is not waived and I agree to be held personally liable in the event that the indicated person, company or association fails to pay for any part or the full amount of the charges. A safe deposit box is available for the protection of guest valuables. The Hotel's liability is limited pursuant to general business law.

Your Signature Service Representative

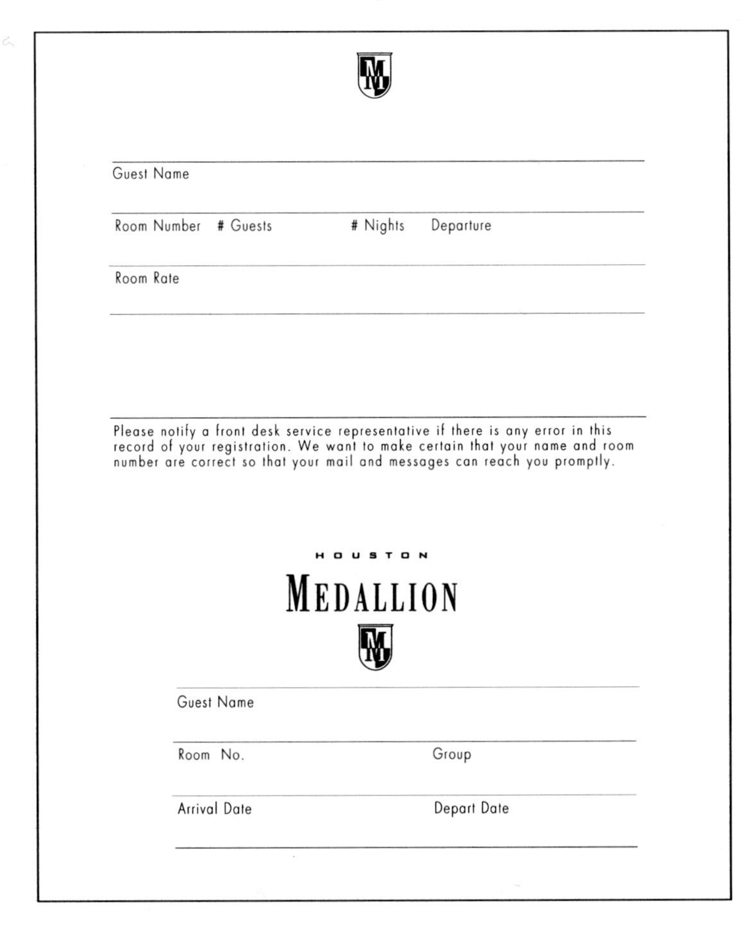

Guest Name

Room Number # Guests # Nights Departure

Room Rate

Please notify a front desk service representative if there is any error in this record of your registration. We want to make certain that your name and room number are correct so that your mail and messages can reach you promptly.

HOUSTON

MEDALLION

Guest Name

Room No. Group

Arrival Date Depart Date

Diana DeLucia Design created a graphics standards manual for implementation of all company materials, including forms and internal items.

shampoo bottles, brochures, checkout lists, signage, and more.

The shampoos were made and filled in Hong Kong and required a two-month lead time. The match boxes were made in Japan and needed a six-month lead. They reluctantly had flip-top matchbooks created for temporary use until the boxes arrived. All together, they produced about 50 different pieces in the three months prior to the opening of the hotel. The client received everything by the time the hotel opened, but it was fast designing and fast producing for Diana DeLucia Design.

Creating Graphic Standards

DeLucia Design also created a graphics standards manual for Medallion Hotels. The manual outlines the design policy and methods of implementation, so any additional materials that are created—whether by DeLucia Design or other contracted firms—conform to the original identity program and present a unified, consistent image. The manual, itself designed on the Macintosh, delineates such items as typefaces, scaling, proportions, proper use of marks and type, and so on.

DeLucia Design receives requests to implement new materials, and refers back to the manual they created. The manual was distributed as hard copy and on disk to the public relations firm in Houston and New York employed by Medallion, and to a local design firm Medallion occasionally uses. It was also intended for internal use so the directors and employees would understand what was required for internally generated items, such as forms, and releases.

The manual is also an excellent reference for identity campaigns for additional hotels that will be added to the group. They can easily adapt the identity of one hotel to another by changing something as simple as the color.

EQUIPMENT

- **Macintosh Quadra 800, Quadra 700, IIcx, IIsi, and Classic families of computers**
- **LaserWriter Pro 600 printer**
- **LaserWriter II laser printer**
- **44Mb SyQuest removable media drives**
- **Quik-Tel modem**
- **Macintosh black-and-white scanner**
- **QuarkXPress**
- **Adobe Illustrator**
- **Aldus FreeHand**
- **Adobe Photoshop**

Some of the staff at Diana DeLucia Design.

CONVERSATIONS

DIANA DELUCIA
President

Diana studied at Parson's School of Design in New York. After acquiring a foundation in design at Anspach Grossman Portugal and Danne Blackburn, she worked at a number of firms and agencies. In 1985, she began Diana DeLucia Design. She is a frequent guest lecturer and symposium participant for such topics as design, ethics, and ecological impact of the design community. Currently, Diana is an Art Director for "Open Dialogue," a documentary about AIDS.

When Macintosh first came out, I was emotionally drawn to it. It was like a new toy. I was working for IBM and I went to a trade show in New Orleans and I saw the Mac there, and it just blew me away because I had never seen something that someone that wasn't computer-savvy could do without a whole lot of learning—you could just use a mouse. It was very exciting. That's when they first started having all the paint programs, so it was a much more creative tool.

I came back and I placed an order and you had to be put on a waiting list because they weren't even out on the market yet. So I had to wait about three months until they came out on the market. That's how it started.

I used it to do a brochure for Chemical Bank. I didn't use it for the typesetting at that time. I just used it for the attendant graphics, which were really just an exercise in being able to play with the various airbrush tools and textures on the computer, and I just used them decoratively.

They were just really textures and patterns that represented different aspects of abstract financial thought. When the Mac first came out, it was only in black-and-white, but I just used the actual laser printout of my Imagewriter printer and I pasted them up on the mechanicals and I specced them for PMS colors.

It was neat because it was one of the first brochures that came out using this technique. I didn't have a whole lot of skill at it, but it was just to easy to make something that wowed everybody, because it was new and you could just kind of play with this new thing, and it made beautiful overlapping shapes and textures.

PATRICIA KOVIC
Creative Director

Patricia earned a BFA from the State University of New York at Buffalo and continued her studies in the MFA program at Brooklyn College. Prior to joining DeLucia Design, she was a designer at McGraw-Hill Publications and Knoll International. She is currently a visiting professor with Pratt Institute's Graduate Communications program, and is an Art Director for "Open Dialogue," a documentary about AIDS.

The biggest challenge with this project was that we had so little time to design a multitude of pieces that would be produced not only by different vendors, but out of different materials.

The computer enabled us to view more variations and more possibilities in far less time. We established a basic foundation of the design system and were then able to build on it for every piece. We determined the logo and typographic treatment as a lock-up, as well as the tracking and kerning and customizing of typefaces. All of this meant an easy call-back for anyone working on the project.

We found that creating everything on the computer was a way to produce all the pieces more efficiently, more accurately, and more consistently.

BARBARA TANIS
Creative Director

Barbara began her career in fine arts before studying graphic design at the University of Cincinnati. She was a Design Director at Muir Cornelius Moore, Donovan and Green, and Bright & Associates in New York. She and Diana worked together at various offices throughout the years, and following a year designing in Switzerland, she returned to New York to re-join DeLucia Design.

When we first started, the only one using the computer was our secretary and office manager. Then we slowly evolved and added computers, and added more, and added scanners and printers. It was really an evolutionary process. We made sure that the people we brought in were skilled, so they helped us evolve as well.

I think it's an important tool to have, an important skill. Everyone that we've interviewed in the past two to three years has been in a school where the computer was a very big part of their education.

I think one thing was the that's made a difference was just the speed of being able to do so many variations plus manipulate things that, before, were either technically impossible or financially cost-prohibitive, such as taking an image and manipulating it through the scanner and changing not only the color but the texture, or even the scale of an image. That was something that was not easily done prior to the computer. Plus the ability now to be able to do prepress things that could often save us money and time down the road, and not have to depend on only an outside service like the printer or separator to do that for us. It gives us a little more freedom and with the technical ability that we have now, I think it expands the possibilities of design.

We don't feel that everything has to be done on the computer. If it's easier and it's better and it saves time—if all the criteria is there, time, money, and it solves the problem—then, of course, we do it. If the problem can't be solved [on the computer], we don't try to make it happen on the computer just because we think it should.

Everyone here, though, still feels it's a tool. It's a great tool, but it's still a tool, and I don't think it really replaces creative ability. I think it enhances it. I think it would be difficult to go back now. Very difficult.

SUMMARY

A corporate identity campaign can be as simple as the creation of a logo and some letterhead, or as complex as signage, uniforms, audio-visual material, promotional gifts, and the corporate name. One of the most obvious examples of the extent to which this can be carried is seen in the hospitality industry, where everything right down to the soap bears the corporate image.

CHAPTER 6

ANNUAL REPORT

This chapter gives you a profile of the conception and production of an annual report—one of the most important vehicles a company has for communicating with its analysts, shareholders, and employees. Learn how Frankfurt Balkind Partners skillfully blended text and fiscal information with hundreds of powerful images to create this powerful, provocative product.

Frankfurt Balkind Partners
Time Wa
VHY
Annual Report
Rapport Annuel
9
Jahresbericht
M E W A R N
Rendiconto Annuale
9
年次報告書
0
Memoria Anual
LEADERSHIP
DIREZIONE
LE LEADERSHIP
NNUAL REPORT
TIMEWAR
リーダーシップ
LIDERAZGO
HBO
FÜHRUNG

Frankfurt Balkind created the first annual report for Time Warner, Inc. in 1989 (cover shown here). This report has been widely acknowledged as a landmark in annual report design. While it was controversial, the report created a new language whose words and pictures were used interchangably. The cover directly and powerfully asks the question on people's minds, "Why did Time and Warner merge?" and the report provides the answer.

INTRODUCTION

Frankfurt Balkind Partners is a Communications Agency of more than 100 people, assembled from eleven countries and is located in New York, Los Angeles and San Francisco. They create many projects for Time Warner, Inc. including its annual report.

Frankfurt Balkind has been greatly influenced by advancing technology. They were a forerunner in computer design, buying their first computer in 1985. The firm has made a serious commitment to keep up with technology, but never letting technology rule its operation. According to Curtis Winslow, Frankfurt Balkind's technology director, the firm doesn't jump to purchase new technology, but by keeping abreast of when a particular technology becomes viable, they ensure that computers are used productively.

Aubrey Balkind, CEO of Frankfurt Balkind, says it is his job to look at the communication problems and opportunities from the clients' perspective, to bring a marketing and communications viewpoint and to assign the right resources of people and technology. He explains:

> *At Frankfurt Balkind we have a full spectrum of communication resources including strategists, marketing planners, creative directors, designers, copy writers, art directors, account directors, technologists, video producers and media specialists and they all need to be able to work together. Some projects require small teams, others span the disciplines.*

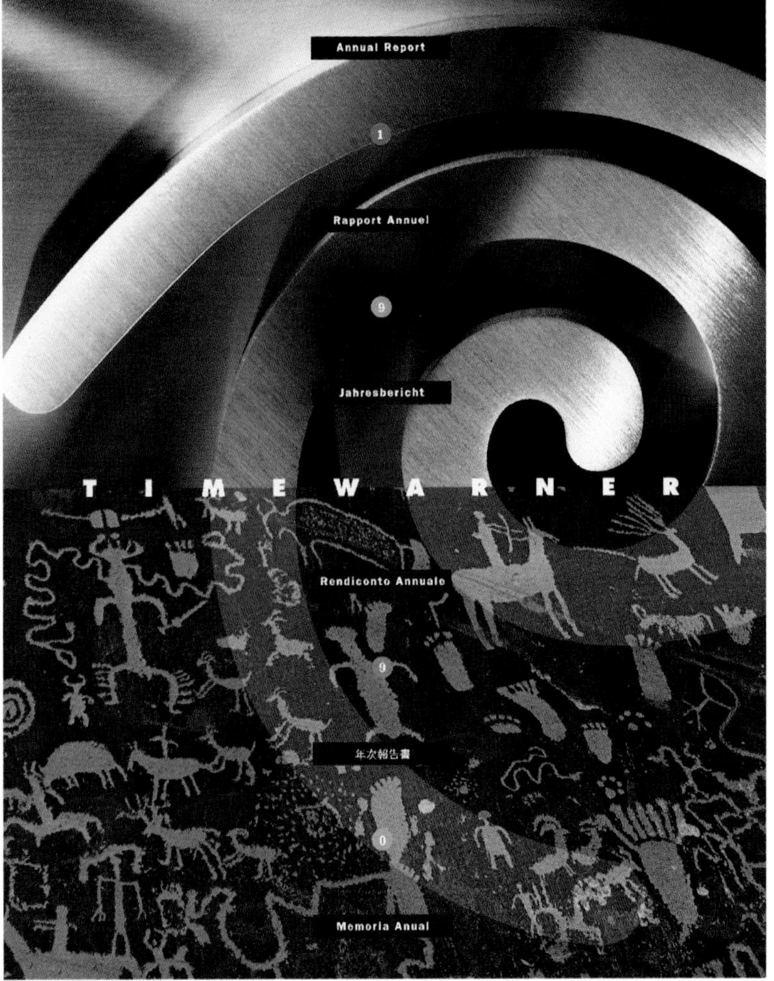

The 1990 Time Warner, Inc. annual report was an interactive report in that the first few pages were slit horizontally, allowing the reader to put different combinations of pages together. Because it was used as a global marketing piece, it was created and produced in six languages.

The 1989 and 1990 Time Warner Annual Reports

The first Time Warner annual report that Frankfurt Balkind created in 1989 is widely considered to be a landmark in annual report design. The problem was how to merge two divergent cultures and competively position them as a new global entity. Working with Time Warner, Frankfurt Balkind made a joint decision to confront head on the question that everyone was asking—why? They answered that and the how, what, and who as well. They agreed to create a new language in which words and images are so closely related that one can be substituted for another. The strategic positioning preceded the design and gave direction to the overall creative approach. While the design was controversial, it took a serious look at how information is communicated in a world of bits and bytes, a world where people scan channels on TV.

Frankfurt Balkind Partners built a new language where words and images worked together, much like a game, giving the book an interactive feel.

The 1990 annual report continued the groundbreaking design. To illustrate vertical and horizontal integration at the company, the first few pages were cut in half horizontally so that the reader could put the pieces of the company together in different combinations. The report was designed and produced in six languages for marketing the company to foreign corporations and governments, and the review section took hypertext to a new level by linking facts, photographs, and graphs with body text.

The cover of the 1991 annual report focuses on Time Warner's products around the world and shows that the report is available in six languages.

Opposite page: To communicate Time Warner's accomplishments, Frankfurt Balkind devised a timeline to highlight the year's events using images from Time Warner's properties, with each chronological spread portraying a different theme. Shown here forming the backdrop for the first four months are images of the Iraq war from CNN and Time *magazine.*

ELEMENTS OF THE PROCESS

When Frankfurt Balkind began the 1991 Time Warner annual report, one of its main goals was to show that Time Warner had accomplished a great many things in the past year. There were a number of major accomplishments that occurred at the company, including some acquisitions, refinancings, product releases, and foreign alliances. Frankfurt Balkind decided to use a timeline to highlight the events of 1991. Says Ruth Diener, the designer in charge of the project:

They [Time Warner] have more than 25 magazines and hundreds of movies and music properties. How do you get all their products into this book in an interesting way? Our solution was to collage images of their products into the timeline and in the introductory pages for each business section.

By using the names of its properties in the business sector collages, they highlighted the fact that each of their businesses is either first or second in its industry. Creating a report for a complex $12 billion company that is in so many interconnected businesses takes planning and strategy. While showing more than 350 images, the report still has a strong focus.

Letter to Shareholders Time Warner is in the business of ideas. Creating them. Shaping them. Reporting them. Inspiring them. Empowering them. Whether it is *Time* magazine's incisive coverage of the momentous events in Eastern Europe; or Time Warner Cable's launch in Queens, New York, of Quantum, the world's first 150-channel cable television system; or the introduction to New York City of its first 24-hour, all-news cable channel; or Warner Bros.' *JFK*, a powerful and controversial treatment of a watershed event in modern history; or Elektra's *Unforgettable* reunion of Natalie Cole and her father; or Gloria Steinem's soulful re-examination of the women's movement in *Revolution From Within*, published by Little, Brown; or HBO's

WARNER BROS. AND THREE EUROPEAN COMPANIES SIGN THE LARGEST-EVER TRANSATLANTIC FINANCING/PRODUCTION/DISTRIBUTION DEAL, VALUED AT $600 MILLION.

LIFE MAGAZINE PUBLISHES WEEKLY EDITION, "*LIFE* IN TIME OF WAR," COVERING THE CONFLICT IN THE PERSIAN GULF.

Gen. Norman Schwarzkopf, Persian Gulf/Life

THE WARNER MUSIC GROUP TOOK OWNERSHIP OF 50 PERCENT OF COLUMBIA HOUSE, OPERATOR OF NORTH AMERICA'S LARGEST MUSIC AND VIDEO CLUBS.

The Josephine Baker Story, the dramatic account of a great entertainer and the racial prejudices she fought against, Time Warner is the leader in bringing audiences here and around the world the very best in information and entertainment. › In all these important efforts, Time Warner depends on one invaluable, irreplaceable resource: our people. Whether they are journalists, photographers, artists, musicians, producers, directors, actors, writers, managers or executives—whatever their role in the Time Warner family—our people are constantly working to broaden the scope of human creativity, understanding and enjoyment. › Supported by the extensive reach of our distribution systems and the growing impact of new technologies, the talent, imagination and commitment of the women and men of Time Warner made 1991 a year of achievement. We announced

2

agreement on a strategic alliance at the subsidiary level. We completed a $2.6 billion rights offering, the largest in history, which strengthened our balance sheet. We pioneered the introduction of multi-channel, interactive cable, a crucial technological innovation in which America leads the world. And none of these achievements came at the expense of our operating results: despite the recession at home and widespread economic uncertainty abroad, four of our five businesses—music, filmed entertainment, HBO and cable—set new records. **World's Leading Creator and Owner of Copyrights** Part of what makes the business of ideas so powerful is copyright ownership. Once an idea takes definitive shape as

THE JOSEPHINE BAKER STORY PREMIERES ON HBO, DRAWING POWERFUL RATINGS AND EVENTUALLY WINNING MULTIPLE CABLEACE AND EMMY AWARDS.

Iraqi soldiers surrendering/*Life*

HBO INDEPENDENT PRODUCTIONS, THE NEW COMEDY PROGRAMMING UNIT, LANDS ITS FIRST PRIMETIME SERIES WHEN FOX BROADCASTING SIGNS *ROC*.

TIME INC. AND GENERAL MOTORS ENTER INTO A MULTI-MILLION-DOLLAR AGREEMENT DESIGNED TO ALLOW INCREASED ACCESS AND MUTUAL INVOLVEMENT IN THE DEVELOPMENT OF CROSS-MEDIA PACKAGES.

TIME AND WARNER NEW MEDIA'S MULTIMEDIA "DESERT STORM" CD-ROM MAKES AVAILABLE ALL OF *TIME'S* GULF WAR COVERAGE, INCLUDING MAPS, PHOTOS AND UNEDITED FIELD CORRESPONDENTS' FILES.

a book, article, CD, album, film, videocassette or TV show—creative software whose copyrights Time Warner owns—it can be sold, licensed and used over and over again, in different formats and languages, through different technologies, to different audiences. The more a company controls not only the creation of these software copyrights but the various forms of their distribution, the greater its potential profits. › Time Warner is both the world's leading creator and owner of software copyrights and the only media and entertainment company to control 100 percent of its distribution. We are unique. In Asia, thanks to the leadership of the U.S. Government, growing adherence to international copyright laws gives us the chance to capture substantial revenues that up to now have been pirated. The eventual introduction of effective copyright protection in Eastern Europe will

3

As a backdrop to the timeline, Frankfurt Balkind used images from Time Warner, from its magazine divisions and from CNN. The images were arranged according to themes: Spreads focus on the war in the Persian Gulf, the arts, Boris Yeltsin in Red Square, the end of South African Apartheid, Clarence Thomas Senate confirmation hearings, Magic Johnson announcing that he is HIV positive, and the release of Terry Anderson. Emanating from the images on the timeline are captions illustrating the many Time Warner events of 1992.

Next comes the review section, which deals with each line of business: publishing, music, film, HBO, and cable. The visual emphasis of this section is on the brand properties held by Time Warner. Each division is presented as a collage of all of their logos mixed with images that identify that division.

Facing Their Biggest Challenge

According to Peter Belsky, Frankfurt Balkind's Director of Electronic Production, the biggest challenge with the Time Warner annual report was the collages.

COLOR PROOFING

To proof color and create color comps, Frankfurt Balkind first used a Canon CLC-500 color laser copier with a Fiery ColorLaser Configurable PostScript Interpreter (CPSI), manufactured by Electronics For Imaging. The Fiery allows anyone on the network to output continuous-tone color prints directly from their computer to the CLC-500. It interprets the PostScript instructions from Photoshop, Illustrator, and QuarkXPress into control codes for the CLC-500.

When the Fiery was released in 1991 it was a PostScript Level I CPSI (PostScript Level 2 color devices are only now reaching the market). Level I CPSI's have had color-matching problems in the past, due to the RGB to CMYK conversion algorithms, which have been greatly improved in Level 2 devices. Blues, for example, tend to print darker than that displayed on the screen. Also, the printing gamut for the CLC-500 is radically different from the gamut for an offset press. Both of these factors do preclude using the Fiery as an accurate final proofing system. It is however, an extremely cost-effective and efficient system for producing color comps on plain paper.

Final proofs for the Time Warner annual report were output to Kodak XL continuous-tone, dye-sublimation printer. There were still some discrepancies with the matchprints, so Frankfurt Balkind Partners had to perform some last minute color corrections.

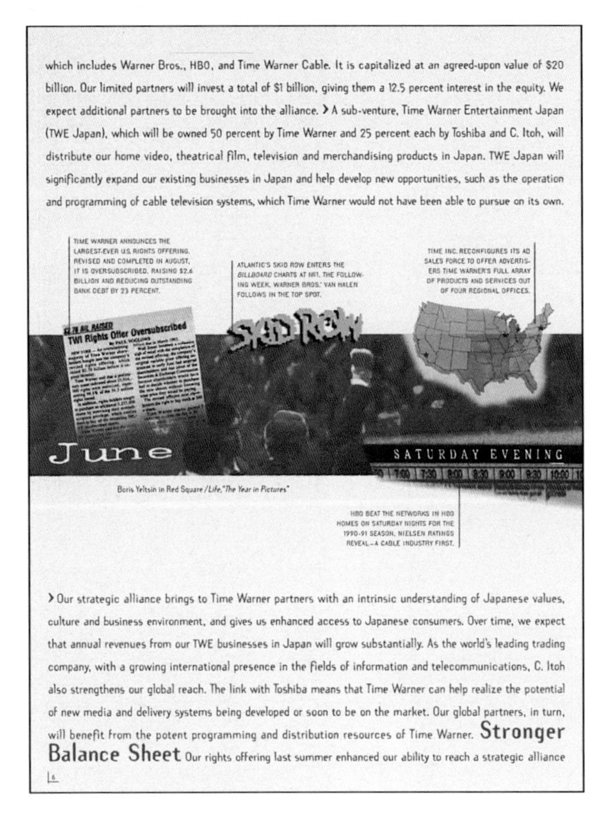

which includes Warner Bros., HBO, and Time Warner Cable. It is capitalized at an agreed-upon value of $20 billion. Our limited partners will invest a total of $1 billion, giving them a 12.5 percent interest in the equity. We expect additional partners to be brought into the alliance. › A sub-venture, Time Warner Entertainment Japan (TWE Japan), which will be owned 50 percent by Time Warner and 25 percent each by Toshiba and C. Itoh, will distribute our home video, theatrical film, television and merchandising products in Japan. TWE Japan will significantly expand our existing businesses in Japan and help develop new opportunities, such as the operation and programming of cable television systems, which Time Warner would not have been able to pursue on its own.

TIME WARNER ANNOUNCES THE LARGEST-EVER U.S. RIGHTS OFFERING. REVISED AND COMPLETED IN AUGUST, IT IS OVERSUBSCRIBED, RAISING $2.6 BILLION AND REDUCING OUTSTANDING BANK DEBT BY 23 PERCENT.

ATLANTIC'S SKID ROW ENTERS THE *BILLBOARD* CHARTS AT NO1. THE FOLLOWING WEEK, WARNER BROS.' VAN HALEN FOLLOWS IN THE TOP SPOT.

TIME INC. RECONFIGURES ITS AD SALES FORCE TO OFFER ADVERTISERS TIME WARNER'S FULL ARRAY OF PRODUCTS AND SERVICES OUT OF FOUR REGIONAL OFFICES.

Boris Yeltsin in Red Square / *Life*, "The Year in Pictures"

HBO BEAT THE NETWORKS IN HBO HOMES ON SATURDAY NIGHTS FOR THE 1990-91 SEASON, NIELSEN RATINGS REVEAL—A CABLE INDUSTRY FIRST.

› Our strategic alliance brings to Time Warner partners with an intrinsic understanding of Japanese values, culture and business environment, and gives us enhanced access to Japanese consumers. Over time, we expect that annual revenues from our TWE businesses in Japan will grow substantially. As the world's leading trading company, with a growing international presence in the fields of information and telecommunications, C. Itoh also strengthens our global reach. The link with Toshiba means that Time Warner can help realize the potential of new media and delivery systems being developed or soon to be on the market. Our global partners, in turn, will benefit from the potent programming and distribution resources of Time Warner. **Stronger Balance Sheet** Our rights offering last summer enhanced our ability to reach a strategic alliance

6

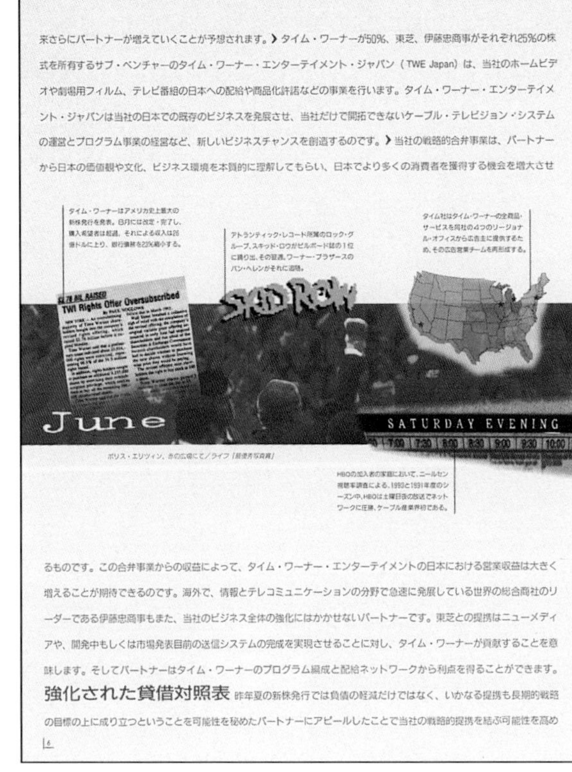

来さらにパートナーが増えていくことが予想されます。›タイム・ワーナーが50%、東芝、伊藤忠商事がそれぞれ25%の株式を所有するサブ・ベンチャーのタイム・ワーナー・エンターテイメント・ジャパン（TWE Japan）は、当社のホームビデオや劇場用フィルム、テレビ番組の日本への配給や商品化許諾などの事業を行います。タイム・ワーナー・エンターテイメント・ジャパンは当社の日本での既存のビジネスを発展させ、当社だけで開拓できないケーブル・テレビジョン・システムの運営とプログラム事業の経営など、新しいビジネスチャンスを創造するのです。›当社の戦略的合弁事業は、パートナーから日本の価値観や文化、ビジネス環境を本質的に理解してもらい、日本でより多くの消費者を獲得する機会を増大させ

タイム・ワーナーはアメリカ史上最大の新株発行を発表。8月には改定・完了し、購入希望者は超過、それによる収入は26億ドルに上り、銀行債務を23%縮小する。

タイム社はタイム・ワーナーの全商品・サービスを同社の4つのリージョナル・オフィスから広告主に提供するため、その広告営業チームを再編成する。

HBOの加入者の家庭において、ニールセン視聴率調査による、1990と1991年度のシーズン中、HBOは土曜日夜の放送でネットワークに圧勝、ケーブル産業界初である。

るものです。この合弁事業からの収益によって、タイム・ワーナー・エンターテイメントの日本における営業収益は大きく増えることが期待できるのです。海外で、情報とテレコミュニケーションの分野で急速に発展している世界の総合商社のリーダーである伊藤忠商事もまた、当社のビジネス全体の強化にはかかせないパートナーです。東芝との提携はニューメディアや、開発中もしくは市場発表目前の送信システムの完成を実現させることに対し、タイム・ワーナーが貢献することを意味します。そしてパートナーはタイム・ワーナーのプログラム編成と配給ネットワークから利点を得ることができます。**強化された貸借対照表** 昨年夏の新株発行では負債の軽減だけではなく、いかなる提携も長期的戦略の目標の上に成り立つということを可能性を秘めたパートナーにアピールしたことで当社の戦略的提携を結ぶ可能性を高め

6

The annual report was printed in five languages. To ease the production, as well as to cut costs, Frankfurt Balkind Partners produced all of the images using four-color process and all of the text using two Pantone Matching System (PMS) spot colors. For all of the versions, the images were printed on a four-color press. Next, separate runs were done for each language version using the PMS colors.

There are about 350 scanned images in the book. Frankfurt Balkind provided artwork to the illustrator, Josh Gosfield. He assembled the images into collages using Adobe Photoshop. Frankfurt Balkind then had the film output at high-resolution and provided it to the printer, so there were no separations required from the printer for the illustrations. Separate materials were given to the printer for the type.

The artwork for the collages was a mixture of conventional scans and video grabs captured from video tape. The video grabs were done with a TrueVision NuVista board which converts any frame of video into an 72 dpi RGB TIFF. The TIFF files were then processed in Adobe Photoshop 2.0. (see the sidebar on Color Proofing).

The style of the illustrator enabled the use of relatively low-resolution scans, which had the added benefit of keeping the file size manageable on the Macintosh. The files were then resampled immediately before they were "ripped," and output as film, which avoided any unwanted pixelization that might have otherwise occurred. The net result was that Frankfurt Balkind was able to work on files that were small enough for efficient work on Macintosh computers without sacrificing quality or creativity.

Using PMS Colors for Multiple-Language Print Runs

The Time Warner annual report had to be printed in six languages. Normally, having six sepa-

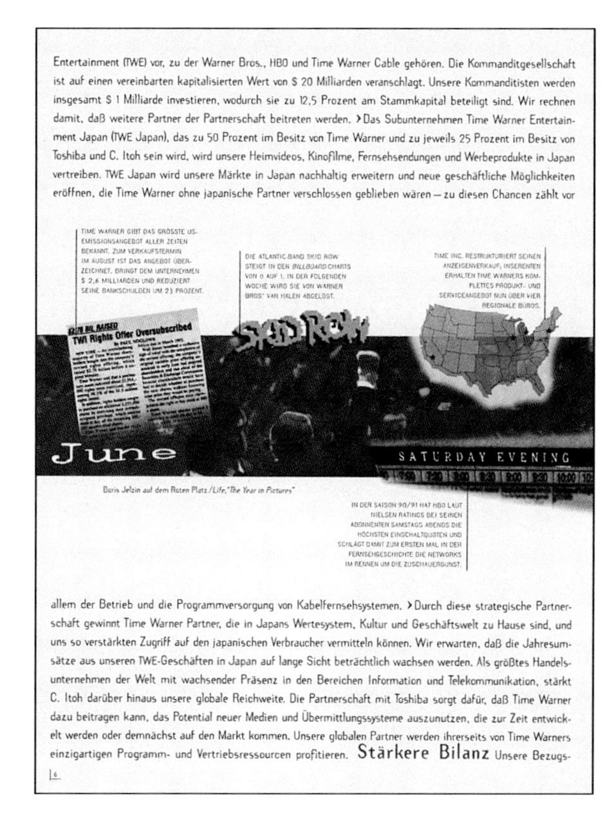

Entertainment (TWE) vor, zu der Warner Bros., HBO und Time Warner Cable gehören. Die Kommanditgesellschaft ist auf einen vereinbarten kapitalisierten Wert von $ 20 Milliarden veranschlagt. Unsere Kommanditisten werden insgesamt $ 1 Milliarde investieren, wodurch sie zu 12,5 Prozent am Stammkapital beteiligt sind. Wir rechnen damit, daß weitere Partner der Partnerschaft beitreten werden. › Das Subunternehmen Time Warner Entertainment Japan (TWE Japan), das zu 50 Prozent im Besitz von Time Warner und zu jeweils 25 Prozent im Besitz von Toshiba und C. Itoh sein wird, wird unsere Heimvideos, Kinofilme, Fernsehsendungen und Werbeprodukte in Japan vertreiben. TWE Japan wird unsere Märkte in Japan nachhaltig erweitern und neue geschäftliche Möglichkeiten eröffnen, die Time Warner ohne japanische Partner verschlossen geblieben wären – zu diesen Chancen zählt vor

TIME WARNER GIBT DAS GRÖSSTE US-EMISSIONSANGEBOT ALLER ZEITEN BEKANNT. ZUM VERKAUFSTERMIN IM AUGUST IST DAS ANGEBOT ÜBERZEICHNET, BRINGT DEM UNTERNEHMEN $ 2,6 MILLIARDEN UND REDUZIERT SEINE BANKSCHULDEN UM 23 PROZENT.

DIE ATLANTIC-BAND SKID ROW STEIGT IN DEN *BILLBOARD* CHARTS VON 0 AUF 1. IN DER FOLGENDEN WOCHE WIRD SIE VON WARNER BROS.' VAN HALEN ABGELÖST.

TIME INC. RESTRUKTURIERT SEINEN ANZEIGENVERKAUF; INSERENTEN ERHALTEN TIME WARNERS KOMPLETTES PRODUKT- UND SERVICEANGEBOT NUN ÜBER VIER REGIONALE BÜROS.

TWI Rights Offer Oversubscribed

SKID ROW

June

SATURDAY EVENING

Boris Jelzin auf dem Roten Platz / *Life*, "The Year in Pictures"

IN DER SAISON 90/91 HAT HBO LAUT NIELSEN RATINGS BEI SEINEN ABONNENTEN SAMSTAGS ABENDS DIE HÖCHSTEN EINSCHALTQUOTEN UND SCHLÄGT DAMIT ZUM ERSTEN MAL IN DER FERNSEHGESCHICHTE DIE NETWORKS IM RENNEN UM DIE ZUSCHAUERGUNST.

allem der Betrieb und die Programmversorgung von Kabelfernsehsystemen. › Durch diese strategische Partnerschaft gewinnt Time Warner Partner, die in Japans Wertesystem, Kultur und Geschäftswelt zu Hause sind, und uns so verstärkten Zugriff auf den japanischen Verbraucher vermitteln können. Wir erwarten, daß die Jahresumsätze aus unseren TWE-Geschäften in Japan auf lange Sicht beträchtlich wachsen werden. Als größtes Handelsunternehmen der Welt mit wachsender Präsenz in den Bereichen Information und Telekommunikation, stärkt C. Itoh darüber hinaus unsere globale Reichweite. Die Partnerschaft mit Toshiba sorgt dafür, daß Time Warner dazu beitragen kann, das Potential neuer Medien und Übermittlungssysteme auszunutzen, die zur Zeit entwickelt werden oder demnächst auf den Markt kommen. Unsere globalen Partner werden ihrerseits von Time Warners einzigartigen Programm- und Vertriebsressourcen profitieren. **Stärkere Bilanz** Unsere Bezugs-

6

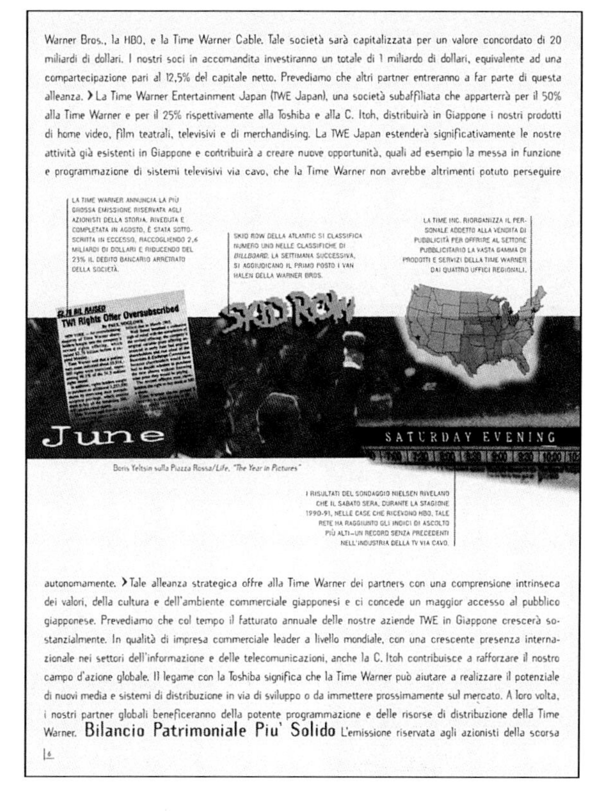

Warner Bros., la HBO, e la Time Warner Cable. Tale società sarà capitalizzata per un valore concordato di 20 miliardi di dollari. I nostri soci in accomandita investiranno un totale di 1 miliardo di dollari, equivalente ad una compartecipazione pari al 12,5% del capitale netto. Prevediamo che altri partner entreranno a far parte di questa alleanza. › La Time Warner Entertainment Japan (TWE Japan), una società subaffiliata che apparterrà per il 50% alla Time Warner e per il 25% rispettivamente alla Toshiba e alla C. Itoh, distribuirà in Giappone i nostri prodotti di home video, film teatrali, televisivi e di merchandising. La TWE Japan estenderà significativamente le nostre attività già esistenti in Giappone e contribuirà a creare nuove opportunità, quali ad esempio la messa in funzione e programmazione di sistemi televisivi via cavo, che la Time Warner non avrebbe altrimenti potuto perseguire

LA TIME WARNER ANNUNCIA LA PIÙ GROSSA EMISSIONE RISERVATA AGLI AZIONISTI DELLA STORIA. RIVEDUTA E COMPLETATA IN AGOSTO, È STATA SOTTOSCRITTA IN ECCESSO, RACCOGLIENDO 2,6 MILIARDI DI DOLLARI E RIDUCENDO DEL 23% IL DEBITO BANCARIO ARRETRATO DELLA SOCIETÀ.

SKID ROW DELLA ATLANTIC SI CLASSIFICA NUMERO UNO NELLE CLASSIFICHE DI *BILLBOARD*. LA SETTIMANA SUCCESSIVA, SI AGGIUDICANO IL PRIMO POSTO I VAN HALEN DELLA WARNER BROS.

LA TIME INC. RIORGANIZZA IL PERSONALE ADDETTO ALLA VENDITA DI PUBBLICITÀ PER OFFRIRE AL SETTORE PUBBLICITARIO LA VASTA GAMMA DI PRODOTTI E SERVIZI DELLA TIME WARNER DAI QUATTRO UFFICI REGIONALI.

TWI Rights Offer Oversubscribed

SKID ROW

June

SATURDAY EVENING

Boris Yeltsin sulla Piazza Rossa/*Life*, "The Year in Pictures"

I RISULTATI DEL SONDAGGIO NIELSEN RIVELANO CHE IL SABATO SERA, DURANTE LA STAGIONE 1990-91, NELLE CASE CHE RICEVONO HBO, TALE RETE HA RAGGIUNTO GLI INDICI DI ASCOLTO PIÙ ALTI–UN RECORD SENZA PRECEDENTI NELL'INDUSTRIA DELLA TV VIA CAVO.

autonomamente. › Tale alleanza strategica offre alla Time Warner dei partners con una comprensione intrinseca dei valori, della cultura e dell'ambiente commerciale giapponesi e ci concede un maggior accesso al pubblico giapponese. Prevediamo che col tempo il fatturato annuale delle nostre aziende TWE in Giappone crescerà sostanzialmente. In qualità di impresa commerciale leader a livello mondiale, con una crescente presenza internazionale nei settori dell'informazione e delle telecomunicazioni, anche la C. Itoh contribuisce a rafforzare il nostro campo d'azione globale. Il legame con la Toshiba significa che la Time Warner può aiutare a realizzare il potenziale di nuovi media e sistemi di distribuzione in via di sviluppo o da immettere prossimamente sul mercato. A loro volta, i nostri partner globali beneficeranno della potente programmazione e delle risorse di distribuzione della Time Warner. **Bilancio Patrimoniale Piu' Solido** L'emissione riservata agli azionisti della scorsa

6

Four of the language versions are shown here: opposite page, English and Japanese; this page, German and Italian.

rate print runs for a product that uses four-color process printing, with two additional PMS colors, is a pricey proposition. But Frankfurt Balkind cleverly designed around that, thanks to an idea from Fred Bard, the Print Production Manager.

All of the images in the annual report were done as four-color. The text was printed in two PMS colors. This allowed them to print all of the four-color work at once, on a web press, with the English text. Later, the other five languages were imprinted using the two PMS colors on sheet-fed presses. In comparison to the previous edition of the annual report, this was a substantial savings, since no textual material had to be re-run in four-color.

An additional advantage to this design was that the images were quickly finalized and sent to be separated. The text, however, needed constant corrections, right up to press time. These corrections only had to be made on the PMS color plates, not on the process color plates.

It is interesting to note that with all of the technological advantages that Japan has, Japanese language technology is far behind western language technology in terms of typeface options and in terms of Macintosh technology. The software available for typesetting in Japanese was so old that they had to cut and paste in the rules to place them.

Using the QuarkXPress style sheets for language translations was invaluable to the production process. The English version was sent to trans-

The annual report used technology in a very innovative way to make it economically feasible to run over 350 images in one report. Images were collected from print and video, then combined in Photoshop by an illustrator using low-resolution (72 dpi) files. The files were resampled before they were ripped, to avoid any unwanted pixelization.

lators with grids and style sheets. Designer Ruth Diener poured the text into her QuarkXPress files, applied her style sheets, and returned the work on a disk. Ruth altered the specs for the varying lengths, and sent proofs back to the translator to check hyphenation.

Creating and Setting the Type

The annual report was set in one typeface: Template Gothic. The designers, however, made some alterations to the original font, including underscores, and renamed it "Timeplate Gothic," for their own use.

Using Altsys Fontographer, they built the underscores into the typeface so they could get different left and right line endings electronically. Throughout the book (even in the financials) rules were positioned to look randomly placed. Once the characters were modified, the designers did extensive kerning in QuarkXPress, the program used to assemble the documents.

Assembling the Final Piece

QuarkXPress was used to merge the various parts of this project—Frankfurt Balkind's preferred production method. The annual report was then sent to the printer to be ripped (processed) through a Scitex Dolev imagesetter. Frankfurt Balkind uses a professional engraver, or its print-

er, to output about 70% of their work. The remaining 30% is output on resin-coated (RC) paper on the Linotronic L300 in their office. Explains Kent Hunter, one of the creative directors on the project:

> *At this point we still prefer that the printer to do the final film separations. You get a much higher resolution and there's more control over color corrections. The printer is still integral to the project—a partner. You need to work with people who really understand and keep up with the technology.*

EQUIPMENT

- **Macintosh Quadra 950 family of computers with 132Mb of RAM and 1200Mb hard drives**
- **Macintosh Quadra 700 family of computers with 20Mb of RAM and 230Mb hard drives**
- **SuperMac/24 PDQ Plus (v 1.292 card) and SuperMac 20" RGB displays**
- **SuperMac Videospigot Pro**
- **NuDesign 128Mb optical drive**
- **RasterOps Video Expander**
- **Truevision Nuvista+**
- **44Mb SyQuest removeable media drives**
- **Sharp UX-450 scanner**
- **QuarkXPress 3.2**
- **Macromind Director 3.1**
- **Adobe Photoshop 2.5**
- **Adobe Illustrator 3.2.1**
- **Adobe Premier 2.0.1**
- **Cinemation 1.0**
- **Aldus Persuasion 2.1**

CONVERSATIONS

AUBREY BALKIND
Chief Executive Officer

Prior to becoming a designer, Aubrey earned a degree in economics in South Africa, an MBA from Columbia University, completed his studies for a Ph.D. in urban design, programmed computers, and worked as a management consultant for Ernst and Young.

Frankfurt Balkind is what we think of as the new breed of agency. It's completely integrated in that there are specialists in identity, advertising, graphic design, 3-D design, multimedia, and technology, and they all work together. Even though we work in three cities—New York, Los Angeles, and San Francisco—we are not structured in separate divisions but work together as one company, sharing clients and projects. We challenge each other by asking silly questions and then we create surprising solutions. There are teams of writers and designers working with strategists and marketing people determining who the audience is, and what their needs are before coming up with a creative solution that pushes the right button.

Not only do we use technology, the technology actually influences the way we think. For example, the 1992 Time Warner annual [report] is printed on uncoated paper, emulating the copies that we get off our Cannon copier. Computers change your vision because instead of looking at C prints as comps, you're looking at color copier art, which is on uncoated paper. You get a different aesthetic and a different feel. We also wanted the report to be on uncoated recycled paper: the softer feel of uncoated paper is in harmony with the recycled aesthetic.

Computers don't manufacture creative concepts. What they do is allow you to experiment, and that gives you more control and cuts down on turnaround time. In order to grow with technology, you have to make a serious commitment to it. It's not simply buying a computer. There is the software (we probably use over 30 programs), the input and output devices,the networks, prepress systems, color control, digital video, video editing and controllers, digital sound mixers, and so on. And it all keeps evolving, so you have to continually upgrade your systems and software. Its like a moving escalator—you commit and get on it.

Communication, like hardware, is also converging. Our TVs will also be our computers and our telephone networks and all this digital information and entertainment needs to be created and designed.

Know who you want to communicate with, what you want to say, how your expression should feel and then create and produce it. And remember it can be great even though its only zeros and ones.

KENT HUNTER
Executive Design Director

Kent directs a team of designers on projects ranging from annual reports and identity programs to video and multimedia presentations. A native Texan, he worked in Nashville, Tennessee before joining Frankfurt Balkind Partners in 1986. He is now a principal of the firm.

Has the computer revolutionized the way we design? Yes and no. We certainly could not have produced a book like the '91 Time Warner Annual Report five years ago.

As we've "grown–up" with computers, our role as designers has evolved. At first we became typesetters, which gave us lots of options and flexibility. But we also inherited all the responsibility for proofreading and precision we took for granted back in the good old days.

Then, just as we were getting used to typesetter's hours, along came programs like Adobe Photoshop, turning us into photo-illustrators. Those of us who have always collaborated with great photographers and illustrators, suddenly found we were able to create interesting images in-house, combining stock photography with type and video.

This project was an excellent opportunity to work with an artist like Josh Gosfield who has gone through his own evolution—from art direction to traditional illustration to computer illustration. We collaborated with him to create the "electronic graffiti" images throughout the book. Since then, several of the designers at Frankfurt Balkind have become quite adept at Photoshop and we're doing more and more of our own artwork for projects.

Back to the issue of type. Technology has certainly allowed a lot of things to be done in the name of "progress" which typographic purists despise. We really hate excessive condensing and stretching of type. It's painful. Adobe's new Multiple Master font technology truly is revolutionary. It retains the beauty of the letterforms while giving the designer flexibility.

I'm not sure what those type purists think of the Time Warner book. A lot of thought really did go into the typography. Ruth is a type fanatic and I believe it shows in her attention to detail. I think the choice of Template Gothic as the only typeface in the book is what makes it so fresh. Barry Deck, who designed the face, told us he never expected to see it used in something as corporate as an annual report. We loved its raw, almost ugly feel. Very immediate. It worked beautifully with the computer illustrations.

The truth is, the design of the book is actually very simple. A timeline bisects the spreads of the shareholder's letter, followed by full page images and large copy introducing each line of business. The type and illustration are what make the book interesting.

RUTH DIENER
Designer

After graduating from the Rhode Island School of Design, Ruth worked at Nesnadny and Schwartz in Cleveland, Ohio for two years. She then moved to New York and joined Frankfurt Balkind in 1990.

In the past couple of years the design process has changed significantly because of the computer. I used to use the Macintosh as a tool for sketching in order to be able to spec type the traditional way. Now I use the computer to execute the entire process—from sketching ideas, to comping, to final typesetting, and ultimately prepress production. Rough sketches get revised and refined in layers, blurring the lines between layout, typesetting and production. The designer is wearing more and more hats. In the past, the designer spec'd type, and sent it out. It could be back the next morning. At Frankfurt Balkind we have a technology department to help with typesetting, but it is very difficult to know when to hand off a project. Preparing documents for electronic mechanicals and using the Mac for prepress is a huge new responsibility for the designer.

It is easy to get carried away with the possibilities for color when you're sitting at the computer—forgetting the limitations of the final execution of a printed piece. In the past, for example, it was expensive to order rub-downs in different colors for comping, which forced the designer to think out the units available on press. Now all the designers at Frankfurt Balkind are connected to the Fiery on the color copier. This is a great tool that allows us to be more experimental with color and see immediate yet unrealistic results. It also makes it easy to be irresponsible in terms of the printability of the final product.

Unfortunately, the computer has made it easier to produce average design. Holding on to a higher standard of typography requires an attention to detail not built into the software. The most glaring examples of bad computer type are default apostrophes and quote marks. In order to use the punctuation designed for the typeface, the designer has to know the proper key combinations. I cringe when I see bastardized quotes. Kerning is also generally much worse than in traditional typesetting. In the past, type houses refined the kerning of their typefaces. It became the secret recipe of each house. I have yet to find the kerning pairs built into the computer faces to be sufficient. Finessing typography is subjective—it is not strictly mechanical—thank goodness.

CURTIS WINSLOW
Technology Director

Curtis oversees all computer equipment issues at Frankfurt Balkind. His responsibilities fall into two areas—keeping Frankfurt Balkind competitive and servicing the computer needs of their clients. Prior to joining Frankfurt Balkind, Curtis was an information manager for a printing company, and a production manager for a division of First Boston.

I insure that we keep a careful balance on the line between "leading edge" and "bleeding edge." Obviously, we don't purchase every new toy that comes along, but we will acquire emerging technologies once they're proven to be reliable and profitable.

In order that the larger system works, and production flows smoothly, a lot of little systems have to be in place first. Font management, for instance. One of the prime reasons that causes an application like QuarkXPress to crash is a corrupt font. Unless the fonts are stored and managed in a consistent way across every designer's machine, then isolating the cause of the crash can take hours, if not days.

Another thing at Frankfurt Balkind is that everybody here is at different speeds in terms of their computer expertise. The technology department provides as much support as possible to make up that expertise difference. I spend a good deal of my time using the software the designers are using, but if I don't know the answer that's application-specific, like a Quark or Photoshop question, then Peter usually does.

I'm also responsible for setting up and maintaining our telecommunications links between ourselves, our clients, and our vendors. We have electronic mail in our office with gateways linking our network to MCI Mail, AppleLink, Compuserve, the Internet, as well as our client's electronic mail systems. When possible we always try to establish and electronic link with our clients. During annual [report] season, we constantly modem files to our service bureaus, printers, or separators. This used to require a dedicated computer and operator to oversee that the modem transmission went through, so we installed a communications server and automated the process. This trend towards electronic communication, as opposed to overnight mail, faxes, and telephone calls, will increase as more and more companies get personal computers. We installed ISDN lines this year to take advantage of the emerging digital phone technologies. The biggest computer and technology issues for the 1990s will be involving connectivity and telecommunications, and I'll make sure Frankfurt Balkind will be ready for it.

PETER BELSKY
Director of Electronic Production

Peter has specialized in electronic production for over six years. Prior to working at Frankfurt Balkind, he worked as a consultant in systems development and electronic production.

A typical day consists of at least one designer coming down and saying "I've got this crazy idea, how can we do it?" That's where the challenge is, and that's what makes our job fun. Additionally, we in the Electronic Production department pick up where the designers leave off, and see the job through to the printer. Different designers are at different levels—at different places with the computer. Some are very, very fluent. Others are not so fluent. Some are incredibly aware of typography. Some don't want to think about typography. For those who don't want to think about typography, we're their typesetters. For those who are very computer literate, we're there to help them see there jobs through, as well as to show them ways to use the computer that they might not have known about before.

We're what traditional production was a few years ago. I still interface with traditional production on certain jobs, but a good 80% of what we do is going out completely electronic.

I troubleshoot. Curtis troubleshoots. Curtis is more into the hardware, software, and the network. I'm more concerned with the software as it relates to the project at hand. Curtis is big picture. I'm more "how do I achieve this effect now?"

SUMMARY

Companies must promote themselves and their products not only to the general marketplace, but to their stockholders, employees and investors as well. An annual report is one of the most important vehicles a company has to present itself to its shareholders. Provocative, powerful images are used to offset textual and financial information, and to "push the emotional buttons" of the readers to achieve a desired response. As explained by Aubrey Balkind, for this to work well, the focus must be on the strategy and the creative product. The creative product must not only push a button, but push the right one, or the wrong result can be achieved.

CHAPTER 7

PACKAGE DESIGN

This chapter follows the creation of ground-breaking packaging for Fractal Design Painter. Learn how this design firm took a mixed metaphor, a sense of humor, and the product itself to create a package that is in as much demand as the product it serves.

Rucker Huggins Design/Fractal Design Painter

INTRODUCTION

Fractal Design Painter is a 24-bit painting and image-editing program for both Macintosh and Microsoft Windows platforms. The program is designed to simulate the tools and textures of natural painting and drawing media.

When the application's development was nearing completion, Fractal Design needed to hire a design firm to create its packaging. They chose Hal Rucker and Cleo Huggins, partners in Rucker Huggins Design in Mountain View, California.

One of the many unique aspects of this project is best said by Stephen Manousos, Vice President in charge of software development:

> *The package we developed was the package for the software we used to develop the package.*

In a certain respect, it's science imitating art imitating science. In another respect, it's a paradox. The Painter application melds fine art with computer technology on level not really seen before. Therefore, so does the creation of the packaging.

In some of the projects discussed in this book, there is a relatively clear line where traditional design and production ends and the technology begins. Here, there's no way to determine where traditional methods leave off and the technology kicks in.

Rucker Huggins presented Fractal Design with six or seven different concepts for the packaging of its Painter application. They presented the paint can concept simply to show they had a sense of humor.

In Painter, any open photograph can be cloned, which means an exact copy is made using the original image as a source (center). After an image is cloned, a cloner tool can be selected—including chalk, pencil, and oil paint—to paint over the image. The image begins to change according to the style of brush chosen. Clockwise from left to right is: hairy cloner, pencil sketch cloner, driving rain cloner, and hard oil cloner.

ABOUT THE PRODUCT

The idea for the Painter application came from Fractal Design president, Mark Zimmer. Mark, with his partner Tom Hedges, had created an earlier program, ImageStudio. ImageStudio developed from an application Mark created for his wife, who enjoyed weaving, to develop patterns for her to weave. From that idea grew ColorStudio, and Painter followed ColorStudio.

Rather than doing photo retouching, which is what ColorStudio is good at, Mark wanted a program that faithfully simulated what can be done with natural tools and textures.

Fractal Design Corp. trademarked a term that best describes what Painter tries to do—*natural media.* Painter simulates natural media on a computer, meaning that it has available all of the tools and the textures that graphic artists would be used to using.

The program is designed for a user who is trained in fine arts, or who knows how to pick up a pencil and draw. The user simply picks up a stylus or the mouse, selects a tool to paint with or draw with, and begins creating. There are no secret buttons to push or special formulas to memorize. It's a very natural program, and the results are very natural.

ELEMENTS OF THE PROCESS

When Hal and Cleo began working on the Painter packaging, Fractal Design had just gotten its Beta (testing) version working. The crew at Fractal Design wasn't quite sure what they wanted, and they were too busy to even think about it. Rucker Huggins decided to present a variety of options to them. Hal describes the presentation:

> *In our first meeting with them we presented them with about six or seven different concepts, ranging*

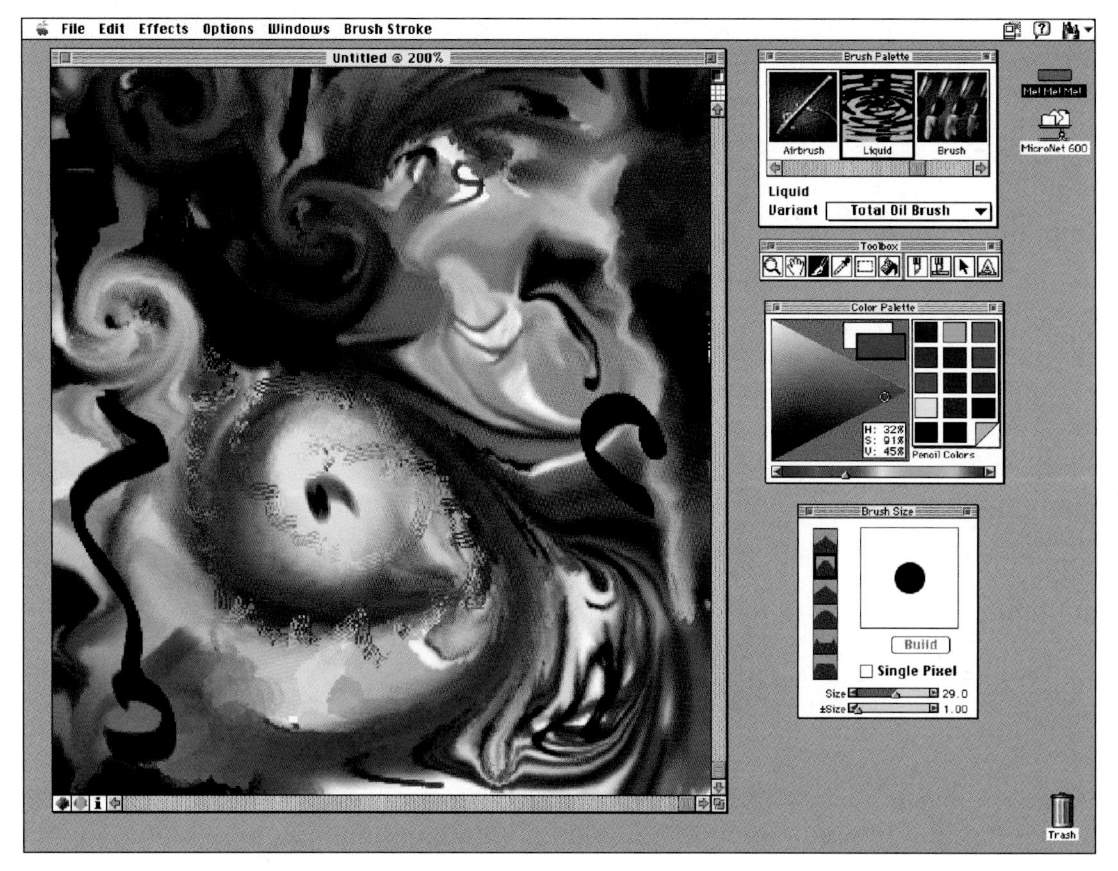

Painter simulates all of the tools and textures available to fine artists. The program is designed so an artist can work as if with traditional art tools: It simply takes a movement of the mouse or stylus to paint or draw. The effects on the screen are natural without the artist having to remember to push any buttons or use any formulas.

from your basic cardboard rectangular solid to some pretty "out there" ideas. It was a pretty interesting meeting. We ended up showing them the paint can just to show them we had a sense of humor. And it wasn't really an idea we liked. They glommed onto the paint can idea and when we showed them my favorite ideas, they weren't interested because they liked the can so much.

It really was just a whim. We'd finished the presentation and I felt it was a little dry, so we tried to think of one more idea. The obvious connection of Painter and a paint can came up.

Rucker Huggins thought it either would be a great success or a complete flop. They felt that the paint can idea was a metaphor that didn't quite work. Fractal Design Painter is a fine-arts application, but the packaging is a housepainter's can: People would think of house painting rather than fine arts painting.

What they decided to do is to really play-up the paint can idea, and make it as inten-

Rucker Huggins also designed the packaging for Fractal Design's Sketcher application. Alternative packaging is sometimes less expensive than traditional packaging: Fractal Design's ColorStudio program is marketed in a traditional box with a sleeve at about $5 per unit; Sketcher's unique packaging costs about $.80.

tional as possible. So they put the picture of the can on the can, and included a poster with cans all over it.

Completing the Concept

After the company decided to use the paint can concept, they had to decide what to do with the label. The most obvious choice was a label that is a long strip that wrapped around the can.

Cleo Huggins then came up with the painter logo, which fit Fractal Design's request for something that looked hand-drawn and free-form.

Then they decided to accentuate the can and take the edge off the mixed metaphor. This was done by reproducing a photograph of a can as an oil painting on the front of the can, with fine-arts tools—brushes—coming out of the can. The result is a picture of a can on a can.

BREAKING THE TRADITION

Rucker Huggins liked the idea of breaking with the traditional software-packaging notion. One of the other initial presentations the design firm made was a little wooden crate. Hal describes the concept:

> *We found a box maker in the Napa Valley making boxes for wines—little wooden crates. And we thought it'd be nice to put it in a wooden crate because that's something we associate with fine arts. We put all our pens and pencils in these old cigar boxes, wooden crates, and things. It ended up we could make those as cheaply as printing the traditional cardboard boxes that you see software in these days. And it has a much higher perceived value than cardboard.*

Fractal Design Painter eventually went back to the wooden box concept for a different product, Sketcher. Sketcher worked out to be an even less costly solution than the Painter one: They simply used a cardboard-like pencil box custom-made by a cigar-box manufacturer.

The biggest challenge of the Sketcher packaging, says Hal, was to develop a concept that complemented the paint can, but at the same time wasn't so gimmicky that people got tired of it.

Fractal Design wanted a hand-drawn, free-form logo. They went to Cleo Huggins, a former Adobe font designer with a Master's degree in typography, to create the logo shown.

On the back of the label are samples of the kind of work that can be done with the product, and some text about its capabilities. On the top of the can is simply a circle with the logo and the platform on which it is used—Macintosh or Microsoft Windows.

The art on the front of the paint can was created by Mark, one of the developers of the Painter software.

To create the art, Mark scanned in a photograph and opened it in Painter. Painter has a cloning process, which makes an exact copy of the image, so that two copies of the image can be open at once. Painter uses the original photograph as the source for the clone. Several parts of the original photograph can be used to achieve a certain look in the clone, including its colors and luminance.

After the image was cloned, Mark chose a cloner tool—for this image, the oil painting tool—then simply painted over the clone. As he works on the image, the colors come from the source, and the brush strokes come from the artist.

Mark also wanted to put in the can a poster of images created in Painter. They talked about folding a poster and placing it inside the can, but didn't like the idea of folding it. Then they

All of the illustration for the packaging was created in Painter, while the product was still being tested. The layout was done on the Macintosh in PageMaker, before Rucker Huggins switched to QuarkXPress. Output was produced by Aptos Post, Aptos, CA, a service bureau associated with Fractal Design.

developed the idea of a long, thin poster that could wrap around the inside of the can. The original poster had five different paint cans on it, all cloned from the same photograph.

The poster shape was chosen because it fit best in the can. However, they also found that it was easy to find a place on the wall to hang that poster—it didn't take up a lot of room. Says Stephen:

We see them all over the place. We see them in windows of computer stores. We've gone into artists' studios and seen them hanging on the wall.

All of the illustration for the package and its contents was done in Painter. The layout was created in Aldus PageMaker, before Rucker Huggins started to use QuarkXPress. The design firm found it very easy to work with Painter, because Fractal Design is associated with Aptos Post, a West coast service bureau. The Aptos Post association took care of all of the usual film output headaches, and eliminated any potential production problems.

One of the challenges for them was the manual, which had to be a unique shape to fit in the can. The paint can has a lot of room inside, but

Mark Zimmer, president of Fractal Design, wanted to include a poster in the can. A traditional-size poster would have to be folded to fit. To avoid this, they developed the idea of a long, thin poster that could be wrapped around the inside of the can.

its shape makes it hard to fit things in it. There was no way to fit a traditional 7" x 9" manual into the can unless it was folded or twisted.

They came up with the lay-flat sketchbook idea, which when opened is long and narrow to mimic the shape of the keyboard. It can be placed on top of or underneath the keyboard, unlike the more cumbersome traditional manual.

Painter is distributed worldwide, but there was a slight hang-up in getting the paint cans in foreign countries. The company did extensive research, but couldn't find one-gallon paint cans in Europe or Asia. In Europe, they had only the imperial gallon, which is a little bigger. So they ship lots of paint cans to Europe, where labels are printed and the packages assembled. In Japan, one distributor

With its lay-flat ring binding, cardboard cover, and textured paper, the manual is designed to look like an artist's sketch book. The design of the manual is also, in part, a result of the dimensions of the paint can: A standard 7" x 9" manual would have to be folded or twisted to fit in the can. When shrink-wrapped with the diskettes, this four-color manual fits snugly in the can without rattling or shifting.

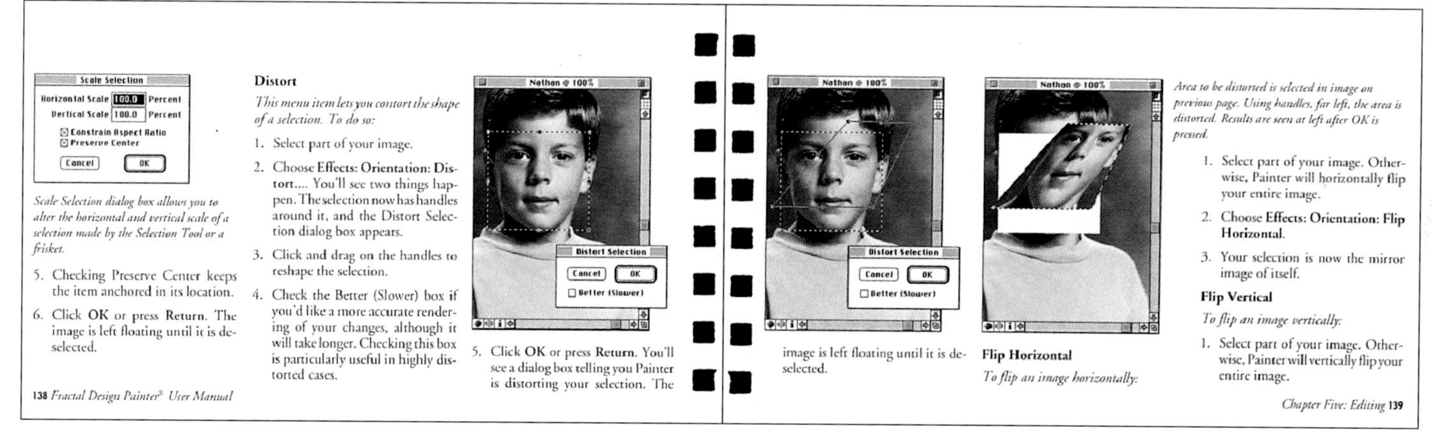

Scale Selection
Horizontal Scale 100.0 Percent
Vertical Scale 100.0 Percent
☒ Constrain Aspect Ratio
☒ Preserve Center
Cancel OK

Scale Selection dialog box allows you to alter the horizontal and vertical scale of a selection made by the Selection Tool or a frisket.

5. Checking Preserve Center keeps the item anchored in its location.
6. Click **OK** or press **Return**. The image is left floating until it is deselected.

Distort

This menu item lets you contort the shape of a selection. To do so:

1. Select part of your image.
2. Choose **Effects: Orientation: Distort....** You'll see two things happen. The selection now has handles around it, and the Distort Selection dialog box appears.
3. Click and drag on the handles to reshape the selection.
4. Check the Better (Slower) box if you'd like a more accurate rendering of your changes, although it will take longer. Checking this box is particularly useful in highly distorted cases.

Nathan @ 100%
Distort Selection
Cancel OK
☐ Better (Slower)

5. Click **OK** or press **Return**. You'll see a dialog box telling you Painter is distorting your selection. The

138 *Fractal Design Painter® User Manual*

Nathan @ 100%
Distort Selection
Cancel OK
☐ Better (Slower)

image is left floating until it is deselected.

Nathan @ 100%

Flip Horizontal

To flip an image horizontally:

Area to be distorted is selected in image on previous page. Using handles, far left, the area is distorted. Results are seen at left after OK is pressed.

1. Select part of your image. Otherwise, Painter will horizontally flip your entire image.
2. Choose **Effects: Orientation: Flip Horizontal.**
3. Your selection is now the mirror image of itself.

Flip Vertical

To flip an image vertically:

1. Select part of your image. Otherwise, Painter will vertically flip your entire image.

Chapter Five: Editing 139

shipped Painter bundled with other products in traditional packaging. The distributor soon received calls from purchasers asking for the paint can.

At about $.72 per can, the packaging is also less expensive than traditional cardboard packaging, which often runs $5 per unit. The cigar box for Sketcher runs about $.80 each. Alternative packages can often be a less expensive solution: Rather than reinventing something, innovative alternative packages often use items widely used elsewhere.

From a logistical point of view, getting paint cans was easy, since the world already makes a million paint cans and there are a lot of vendors available. In fact, the paint can turned out to be a really inexpensive solution.

During the production process, however, it was initially difficult to get the labels to adhere to the cans. The company that produces the labels for Fractal Design was very experienced in traditional software packaging, but the label company had never before dealt with gluing labels on paint cans. Hal explains what happened with the first few batches:

The complete package includes a paint can with label (back right), brochure (back left), manual (front right), diskettes (front left), and poster (page 120).

I got some angry phone calls from them saying "These things won't stick. Our entire office is sticky . . . ." But they obviously eventually figured it out.

Creating an Identity

The paint can added significant product recognition to Painter, and became a great marketing tool for the application. More often than not, customers ask for the paint program that comes in the can. Stephen explains:

A lot of times you'll ask someone if they've ever heard of Fractal Design Painter, and they'll say no. Have you ever heard of Painter? They'll say no. But then you'll ask them if they've heard of the software that comes in the paint can, they'll say they know what that is.

I think it's sort of an icon now that when people see it they think of our software and they think of our company. We use it in all of our ads and all of our brochures.

The response has been overwhelming: The customers love it. Initially, a few people complained that it's not ecologically sound because you can't recycle it. But that's not true, says Stephen, the package is 100% recyclable, in more than one sense.

It can be recycled like you would any can, or you can use it to put things in. You can stick it in the garage and put nails in it, my wife has one with flowers in it on a table in our house. You can put pencils in it. So you can recycle it that way.

EQUIPMENT

- Macintosh IIci with 20Mb RAM and a 240Mb hard drive.
- 21" Apple monitor
- 44Mb SyQuest removable cartridge
- 24-bit RasterOps color card
- Fractal Design Painter
- Fractal Design ColorStudio
- Aldus PageMaker
- QuarkXPress
- Microsoft Word

CONVERSATIONS

STEPHEN MANOUSOS
Vice President of Fractal Design Corp.

Stephen has owned Aptos Post, Inc., a service bureau in Aptos, California, since 1981. Aptos Post was the first service bureau west of the Mississippi, which attracted a lot of developers writing software for desktop publishers who needed a place to test their product. It was there he met the four men he would become partners with in Fractal Design Corp.: Mark Zimmer and Tom Hedges, who developed ColorStudio and ImageStudio; Lee Lorenzen, one of the three developers of Ventura Publisher; and Lee's partner at Altura Software, Steve Thomas.

The kind of software that we have dictated how innovative the packaging should be. We knew it would be used primarily by graphic artists, illustrators, designers, and we wanted our packaging, our manuals, the whole look of the company to appeal to that industry. It's sort of like a typesetter doing his own business card, you want it to look perfect. It just happens that your trade is what is portrayed on that card, unlike other industries. An engineer wouldn't necessarily worry about typography because he's not handing his card to someone who knows the difference. But in our industry we're selling software to people who know the difference and we really want to appeal to them. So we wanted something a little different, we wanted something artsy, something unique. Those were the things we were thinking of when we came up with the paint can.

I think that in the beginning we had some mild objections. But the software started selling so well that those were put to the side. I even heard that in Japan when the software was sold as a bundle with hardware and they bundled it without the can, people complained that they wanted the can and called our distributor trying to get the can after the fact. Sort of a status symbol, I guess.

HAL RUCKER
Rucker Huggins Design

Hal has Bachelor of Science and Master of Science degrees in product design from Stanford University. In addition to working for many years in design, he has made several award-winning animated films. His latest project, Manic Denial, received over a dozen awards, including a first-place award at the San Francisco International Film Festival. Hal's work has been in the Museum of Modern Art in New York, and has aired on national television.

I didn't get into graphic design until computers were around. So I know why I like computers, but I don't know why they're better than before because I wasn't doing it before.

I think graphic design is changing in the sense that you used to have graphic designers and they would hire typographers and an illustrator and their job was to get everything put together. In our office the way things are working now is the designer does everything. If we need an illustrator, the designer does it. If we need a photograph, we shoot it and put it in the computer. We do the typography ourselves. I like that because it seems like it's blurring the tasks. People are treating type as illustration now, people are treating illustration as type, and when one brain is working on the whole project and making all the decisions about type, illustration and all, that I think you get some pretty nice stuff out of it.

CLEO HUGGINS
Rucker Huggins Design

Cleo earned a degree in graphic design from the Rhode Island School of Design, and a Masters of Science degree in digital typography from the Stanford University Department of Computer Science. While at Adobe Systems, she designed Sonata, a musical notation font. She received awards from the New York Type Director's Club for other Adobe Projects. Cleo's work has been featured in several magazines and books about digital design.

We came up with a lot of concepts at the beginning. We wanted to present a range of ideas. We asked questions like "Who's the audience? What's their world like? Where do they keep their paint supplies?" That was one end of the idea—to show it as a place to keep your tools, since they were selling an artistic tool.

This was Hal's and my very first project together. Maybe it even made our business possible because it was our first project. If we had started out doing a nice corporate report, we would have ended up with corporate report projects. But you tend to end up doing work that people have concretely seen. Being allowed to do something as bizarre as the paint can set the tone for allowing us to do more fun stuff.

The best stuff seems to have a sense of humor. Sometimes you get clients who are willing to risk it. If you have a sense of humor on all levels, you can't loose. Those are the most fun pieces.

SUMMARY

We've reached a point in the evolution of the industry where graphic designers are becoming less art directors and more responsible for every level of a project. It remains to be seen if people in the industry are up to the task. The impact of the technology on a project like this is that the lines are being blurred. Designers haven't thrown away traditional approaches, but they are now finishing projects on the computer, because that's the final product.

CHAPTER 8

BROCHURE

This chapter discusses how Ben & Jerry's produces a brochure for its franchise stores. Learn how this company combines its folksy, hand-drawn style with computer technology.

BEN & JERRY'S

Ben & Jerry's has a very identifiable style that is becoming somewhat of a cultural icon.

INTRODUCTION

Most of us are familiar with the mouth-watering taste of Ben & Jerry's ice cream, as well as the company's unique and folksy packaging and displays. The hand-drawn character of their artwork and type, their vivid use of color, as well as their style of insetting illustrations in paragraphs, makes their work (and their product) quite memorable. Their work spans almost all printed media, including billboards, public support material, brochures, banners, and packaging. They even design the interiors of their franchised scoop shops and the signage, as well as their delivery vans.

Their art department uses both traditional and digital production methods, although they sometimes are reluctant to embrace digital tasks beyond computer typesetting, since the nature of their artwork is hand-drawn. Even the digitally created proprietary typeface (seen on every Ben & Jerry's pint of ice cream) has a hand-drawn look and feel to it.

For the most part, Ben & Jerry's mailed pieces and large posters are created traditionally, with only the typesetting done by computer. Some of their brochures and smaller posters, and all of their pint packaging, is created on the computer.

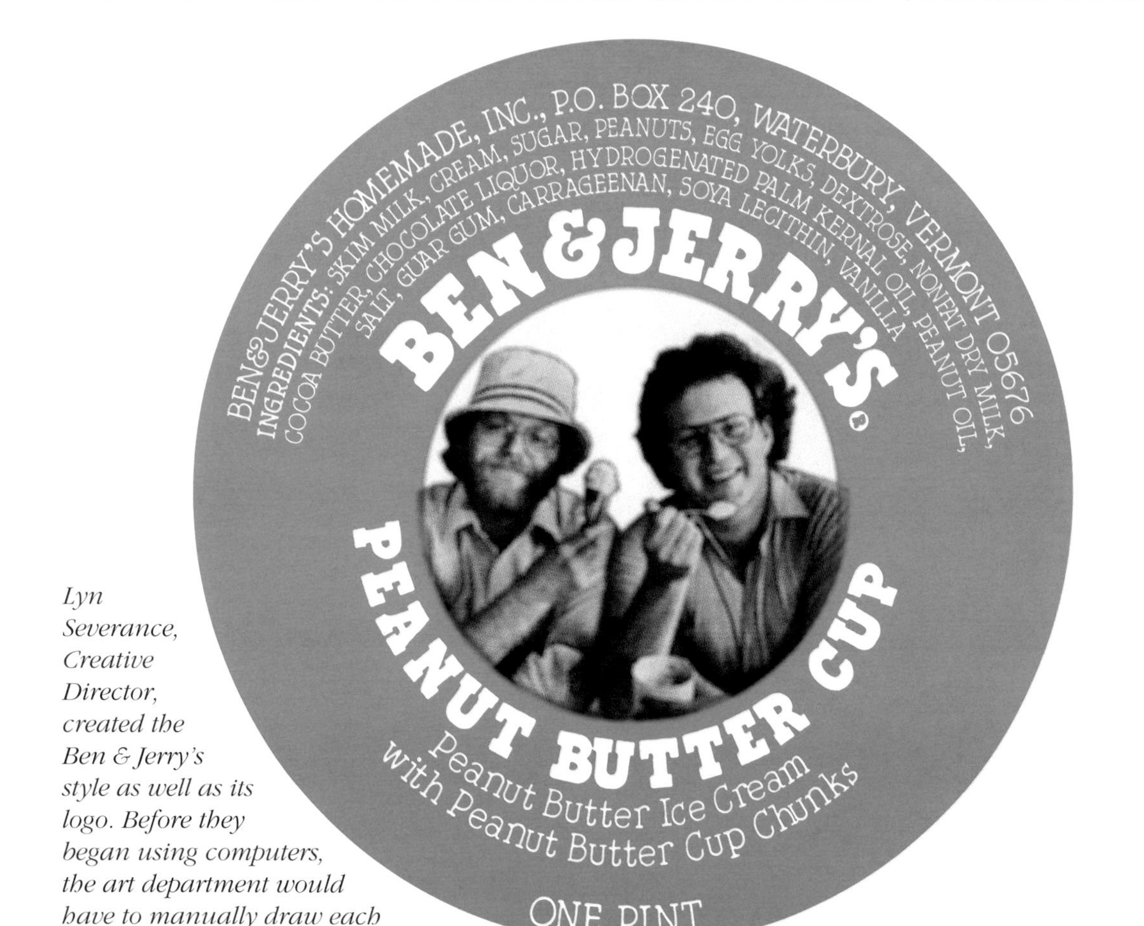

Lyn Severance, Creative Director, created the Ben & Jerry's style as well as its logo. Before they began using computers, the art department would have to manually draw each letter on a circular mechanical.

CREATING THE BEN & JERRY'S STYLE

The Ben & Jerry's style was created by Lyn Severance, who has been working with Ben & Jerry's since its inception. Lyn's artwork has a native folk style, with line drawings and bright colors. Because Ben & Jerry's ice cream is associated with this timeless style, all of the designers who have come into the Ben & Jerry's art department learned to produce material in that same style to produce a consistent visual product.

The Ben & Jerry's logo was created by Lynn during the early days of the company. It may have gotten a little more sophisticated over the years, but it consistently has had a hand-drawn look—no one would ever think of adding a straight line to it.

The need to maintain the folksy, hand-drawn quality in both the logo and the artwork

FONT CREATION

Melissa thinks that one of the reasons her department has been able to use the computer as efficiently as they do is because of the availability of font creation programs. About three and a half years ago, Lynn Severance, now Creative Director, rendered the two most commonly recognizable Ben & Jerry's fonts—Chunk Style and Severance—on Ben & Jerry's first graphics station. They have used these fonts for about 90% of their work ever since, and two years ago added workstations because of the demand.

The seven designers at Ben & Jerry's have all learned to draw in a manner indiscernable from Lyn's.

produced by Ben & Jerry's makes some of the graphic designers hesitant about using digital methods. Sarah Forbes, one of Ben & Jerry's seven graphic designers, explains:

> *Basically we're all learning Lyn's style and perfecting it over the years. This is just her personal style that's very simple, line drawings, bright colors, kind of native folk style. And that was just the logo, Ben and Jerry's, ever since day one. It's certainly gotten a little bit more sophisticated over the years, but with seven of us here creating the style, I think we do a pretty good job of staying consistent. So it's really important that we have it look as if it's done by hand. Therefore there are little mistakes in it, we don't use straight lines ever. Every line is hand drawn and everything's a little bit wiggly. So because of that, it's just about impossible to get that same feel on the computer. At least with the programs that we have currently. The com-*

puter just makes things a little bit too stiff for the feel that we like to have.

ELEMENTS OF THE PROCESS: CREATING A BROCHURE

Ben & Jerry's recently created a brochure for its San Diego area retail shops, titled "The Story of Ben & Jerry's." The brochure contains information about Ben & Jerry's, it's social attitude, product infor-

THE STORY OF BEN&JERRY'S®
VERMONT'S FINEST•ICE CREAM & FROZEN YOGURT™

It all began years ago when two 7th graders became best friends at a gym class in Merrick, Long Island. Neither Ben nor Jerry cared about running a 7-minute mile, but both were crazy for good food. Years later in 1978, by then qualified graduates of a $5.00 ice cream correspondence course, they took over a dilapitated gas station in downtown Burlington, Vermont, and began making homemade ice cream with bigger, more plentiful chunks of good things in it than any other ice cream had ever dared.

Today, Ben & Jerry's makes Vermont's Finest Ice Cream and Frozen Yogurt. We use only superior quality natural ingredients starting with pure Vermont cream and milk, and we add hefty chunks of fruits, nuts, candies, and cookies ... anything that makes for a good adventure and a fine recipe. We're committed to making great ice cream and frozen yogurt and we want you to feel you're treating yourself extra well every time you taste our sundaes, toppings, cones, cakes, shakes and fresh baked goods.

Designer Melissa Salengo chose to use the computer for this brochure because of the need to wrap type around many graphic elements.

mation, and a map and listing for its San Diego area shops. The piece is an 11" x 8.5", four-color piece printed on both sides, that folds down to letter size (3.5" x 8.5").

Creating the Initial Composite

The brochure was designed and executed by Melissa Salengo, a full-time graphic designer at Ben & Jerry's. Initially, Melissa created the rough artwork and preliminary layout by hand—stan-

All art for this brochure was scanned in, then traced using FreeHand.

dard procedure for her department. Melissa decided to use the computer for this job simply because she needed to manipulate the type and because Ben & Jerry's intended to release the project to their printer on film. Otherwise, she would not have opted to create the brochure on the computer.

After creating the initial comps, Melissa reviewed the existing art available, and selected a few images that fit the piece. She created the remainder of the images specifically for this job. Most of the illustrations she created were initially done by hand, then scanned into her Macintosh. She then traced the scans in Aldus FreeHand, and began laying out the brochure in FreeHand. Once completed, the brochure was delivered on disk to their printer.

Sarah Forbes, one of the art directors at Ben & Jerry's, thinks the use of the computer for this project resulted in a better product. She says this type of brochure is an excellent format for the computer,

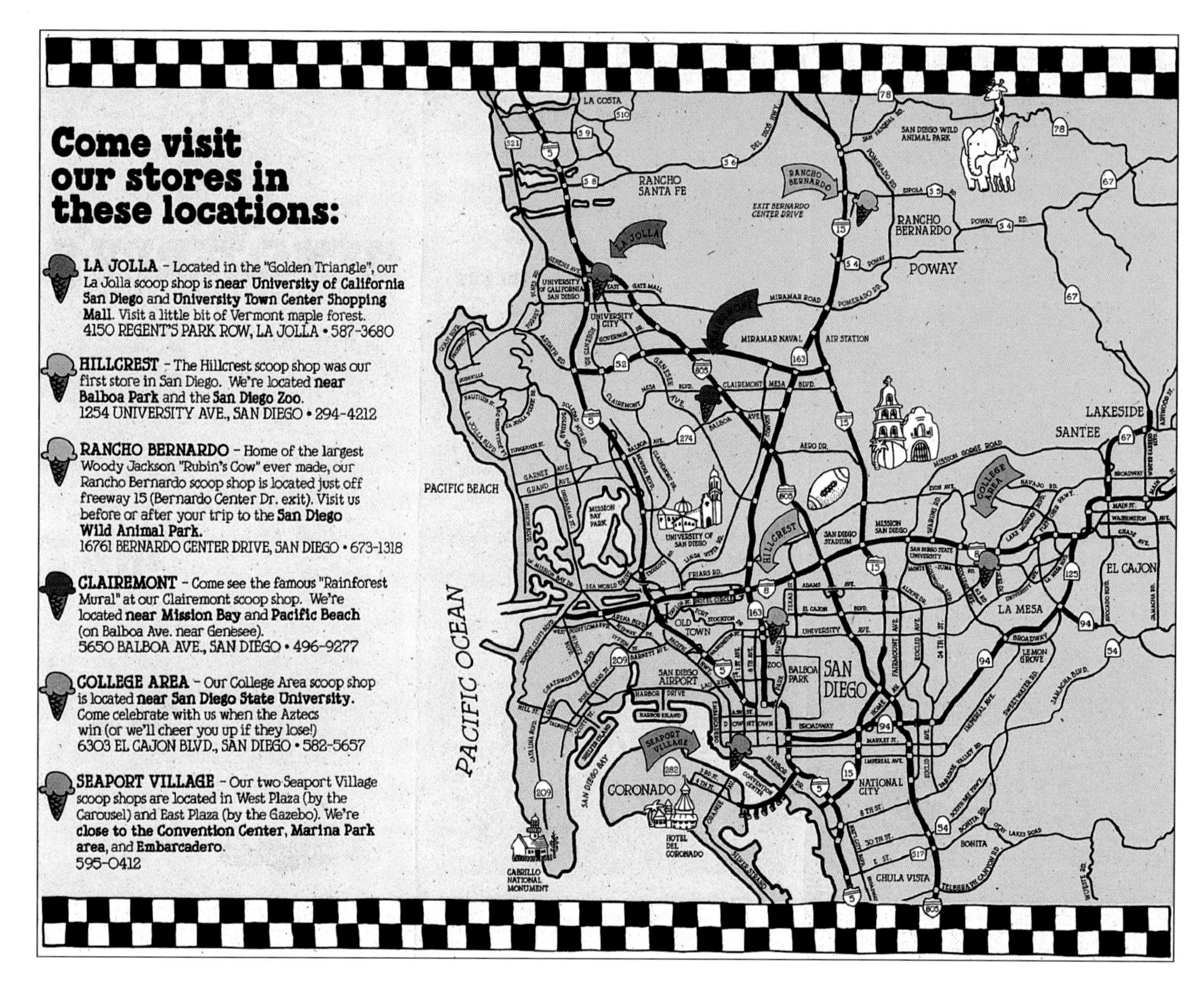

Ben & Jerry's uses four-color printing to emulate the look of a hand-painted watercolor.

since placing illustrations within blocks of text would be cumbersome and very time-consuming if done manually.

Using Process Color for a PMS Color Effect

One of the most identifiable aspects of the folksy Ben & Jerry's artwork is their use of color. Each one of their products, including "The Story of Ben & Jerry's" brochure, uses Pantone (PMS) colors.

For the brochure, those colors were generated on a Macintosh. Melissa culled out all of the colors on the computer, using PMS colors created by using a four-color process. They then used the colors they created with screens, such as 30% magenta or 10% yellow, to create their palette. According to Ben Nooney, Art Department Manager, they can get the result of 30 to 40 PMS colors with this method, which very effectively emulates the look of a hand-painted watercolor.

Ben & Jerry's used Fontographer to create its two most-used proprietary typefaces: Chunk Style and Severance. Until their use of computers, Ben & Jerry's produced by hand all of the artwork and type for their pints.

Designing a Pint

Prior to the availability of digital means, Ben & Jerry's had a very traditional method for producing the curved type found on each one of their pints: They had a curved mechanical and drew all of the type on it by hand. They have since changed to a program, Curve-It, that handles the typographic manipulation for them. They now simply create their art in FreeHand, then curve the type in Curve-It, which curves everything on the right angle and generates the correct proportions. Sarah Forbes explains how this has improved their efficiency:

> *Initially, when we did the pints we did everything by hand, even including all of the type with the decks and the ingredients, and our story on the back with illustrations inserted. With our yogurt line we have a series of illustrations that are within the big oval that is our logo. And we used to do that all by hand, and if there was a typo or a change in the ingredient deck, we would have to go and re-ink the entire thing, unless we could finagle it somehow. It was an incredi-*

bly time-consuming and frustrating, and an amazingly time-intensive process.

Then once we started to get the computers up, we realized how much time and energy it would save us to just be able to have all this on the computer and just have it in there and make the changes quickly, and a lot of times just modem or send the disks to different places to have it printed out. We've almost gotten away from doing mechanicals, now that we've gone to using this Curve-It process.

Sweetheart cups prints all of their pints for them, so they now simply create all of the artwork on disk and send it to Sweetheart.

EQUIPMENT

- **Macintosh II and Quadra families of computers**
- **Radius 2-page color monitors**
- **Aldus FreeHand**
- **Aldus PageMaker**
- **Adobe Photoshop**
- **Curve-It**
- **Altsys Fontographer**

Most things produced by Ben & Jerry's—from ice cream to brochures and packaging—must conform to social and environmental criteria.

ENVIRONMENTAL AND SOCIAL CONSIDERATIONS

One of the most prevalent issues regarding anything produced by Ben & Jerry's—brochures, packaging, ice cream, etc.—is the use and consideration of socially responsible techniques. All of the staff we spoke with placed this as their first consideration, and find it incomprehensible that the rest of the world may operate otherwise.

Recycling

The majority of Ben & Jerry's printed material, except for its food packaging, are printed on recycled paper stock. Although they would like to use recycled materials for those as well, the pints cannot be on recycled paper because it hasn't yet been developed. The largest barrier for that is the need for FDA approval, since it relates to a food product. They have also had trouble finding recycled styrene for displays, as well as recycled banner material.

The Words They Live By

In 1988, Ben & Jerry's Homemade, Inc. created a document called the "Statement of Mission." The use of recycled products is a direct result of this mission, as is the effort to keep a healthy and nonpolluted environment for their artists. They consider the mission to be the cornerstone of their company, and it is quoted here in its entirety.

Ben & Jerry's is dedicated to the creation and demonstration of a new corporate concept of linked prosperity. Our mission consists of three interrelated parts:

Product Mission—To make, distribute, and sell the finest quality, all-natural ice cream and related products in a wide variety

Unless restricted by the FDA, all of Ben & Jerry's printed material is printed on recycled material.

of innovative flavors made from Vermont dairy products.

Social Mission—To operate the company in a way that actively recognizes the central role that business plays in the structure of society by initiating innovative ways to improve the quality of life of a broad community—local, national, and international.

Economic Mission—To operate the company on a sound financial basis of profitable growth, increasing value for our shareholders, and creating career opportunities and financial rewards for our employees.

Underlying the mission of Ben & Jerry's is the determination to seek new and creative ways of addressing all three parts, while holding a deep respect for the individuals, inside and outside the company, and for the communities of which they are a part.

The Ben & Jerry's Foundation

In keeping with their mission, Ben & Jerry's donates 7.5% of its pretax earnings to the Ben & Jerry's Foundation, a nonprofit institution established in 1985 by personal contributions from Ben & Jerry founders Ben Cohen and Jerry Greenfield. The foundation awards monies to nonprofit and charitable organizations through a grant application process and supports "projects which are models for social change; projects infused with a spirit of generosity and hopefulness; projects which enhance people's quality of life; and projects which exhibit creative problem solving."

CONVERSATIONS

SARAH FORBES
Assistant Art Director

Actually I'm also the coordinator for the Green Team, an employee volunteer group—an environmental group within the company. Each location has one. I think the Art Department has three members of the Green Team, and Ben Nooney is really hip to what is happening in the paper industry. I think the Art Department is the most serious about our environmental ethic here. We're very conscious of using recycled papers, with the most consumer content, inks, make sure they're soy-based inks for as much as we can, all the different stuff, the chemicals that are used in the dark rooms and as far as spraying glue, we try to cut out the nasty stuff and use other materials so we can keep this really safe and healthy environment for the artists—keep it non-polluted.

I think we're very successful. We really think each thing through and it's so much a part of the process. Like you have to choose a paper when you design something, it's going to be printed. You have to think about the environmental part of it as easily as that. It just fits in with our process.

I think basically, it stems from our social mission. We do have a three-part social mission that includes an economic mission, a product mission, and a social mission. We're really dedicated the highest quality product that we can produce. And because we're a company, we have stockholders, we need to make money so that we can fulfill our obligation to them. And the third part, which is the most unusual part for most companies, is that we have a social mission which talks about business' responsibility to the community, to give back to the community that supports them, make a nice place for the employees to work, and just to share in that community. The small community that we live in, and also the larger community. So that's one of the things that make us stick out a little bit more than other companies do.

BEN NOONEY
Production Manager

I coordinate all the jobs and schedule them in-house. I coordinate the printing of all of them and all the production. When to get them, where they go, all that good stuff.

I'm an old printer. My background is printing. The problem I have with computers is lack of mechanicals. You're really not sure what you're getting a lot of times until it's already at the printer's and more costly to correct. It's harder to visualize. I'm adjusting to them. We'll leave it at that. I know there are some applications for computers. We're trying to find the balance.

I was working for a printer three to four years ago, before I worked here. Ben & Jerry's was one of my customers just when all this recycled paper was starting to come around. Ben & Jerry's was one of the few companies who would bite the bullet, regardless of the price of the paper because recycled paper was probably 50% to 75% higher than conventional paper. Coworkers would say, "What are those guys crazy? Look how much money they're spending on this." But Ben & Jerry's and other companies like them had the foresight to say "Hey, this is a happening thing." Now it's really in style to use recycled paper and the paper industry has responded by making that paper affordable to everybody. Some smaller companies that didn't have the money, can now afford to do the right thing and use recycled paper and recyclable paper.

I don't think we're unusual. Well, we're unusual in how everyone else perceives us. But it's all common sense to us.

MELISSA SALENGO
Graphic Designer

I think this is kind of a theme that runs throughout. It is that computers are wonderful because we can do the work quickly and we can make changes easily when we used to hand draw all of our type. We had to change a block of type because of a typo, you had to do the whole thing over again. Because of things like that, the computer is just a great thing. Actually almost all of our type, I would say 95% of our type that is used, is generated in the computer. We don't use much hand-drawn type anymore. Pretty much just for headlines and subheads or a lot of the franchise support stuff is done free-hand, not in FreeHand, but done by hand.

SUMMARY

Of all of the companies interviewed for this book, Ben & Jerry's is by far the most reluctant to embrace this new computer technology, although doing so has in many respects been a benefit to them. It is also interesting to note that although the company digitizes some of its images, it only does so by recreating on the computer art created by hand.

CHAPTER 9

LARGE FORMAT DIGITAL OUTPUT

This chapter gives you a profile of the changing world of large format digital output in the advertising and art industry. Learn how this high end fine art print-making studio produces and outputs limited edition exhibition prints and other museum quality visuals for clients.

The Color Space

INTRODUCTION

Jody Dole, Partner in The Color Space, Communications Inc., is a classically trained artist who produces photographs—with traditional camera and film—of a wide array of subjects ranging from liquor advertising to studies of rare animal skeletons. His images are then processed using new digital technology processes that combine photography with advanced computer imaging to create large-scale exhibition prints of exceptional quality. The process revolutionizes photography and imaging by adding an entirely new dimension to the creative image-making process, impacting the art world and a variety of businesses, from advertising and corporate graphics to the entertainment industry.

Jody is a fine art advertising photographer who creates images mainly in the still life area. Although much of his high-level art is limited to galleries and private collections, it is quite likely that you have come across some of his influence every day: for example, the award-winning Smirnoff ads that began his commercial career.

Jody, along with partner Peter X. Ksiezopolski, Creative Director of Peter X(+C) Ltd. Designers, outputs museum-quality prints and separation films in their Manhattan facility, called The Color Space. X(+C) Design was the first New York

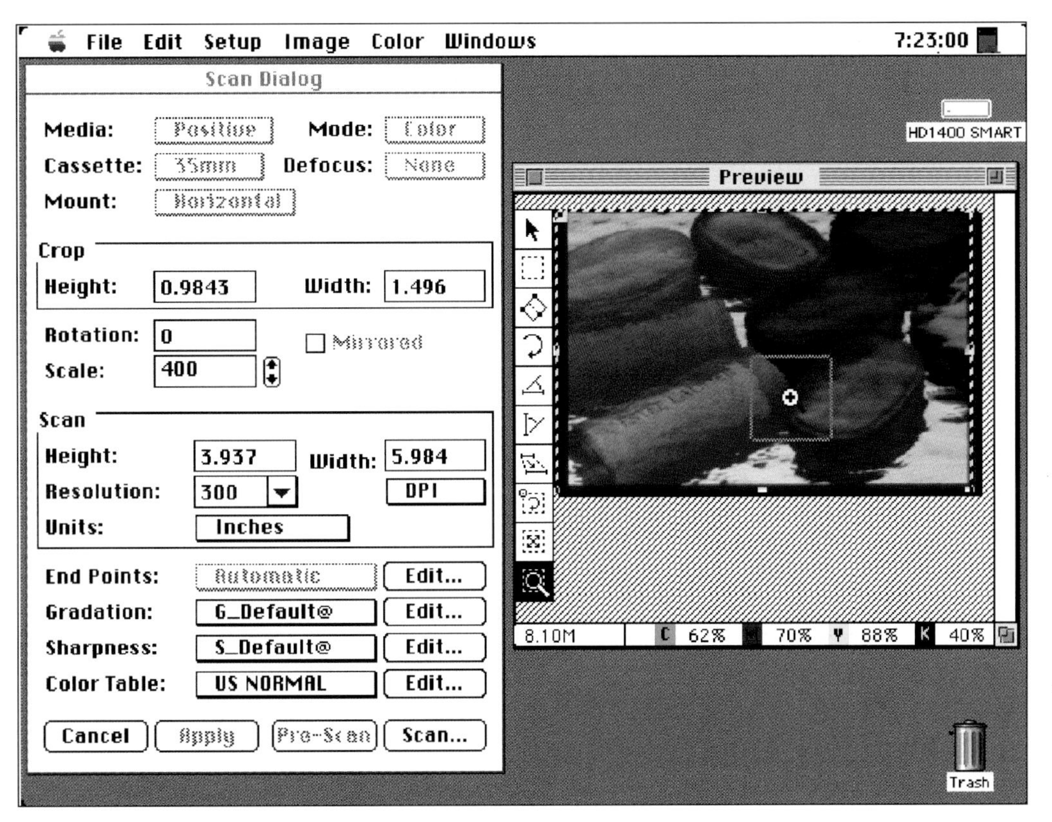

In the pre-scan mode they determine original type and format. They set the resolution sand scale (file size is diploid) and also rotate or crop the image as desired. When rotation is required, several tools are available including selection of two points, which then invoke the scanner carriage to physically rotate. This not only saves time compared to rotating an image in Photoshop, but also avoids the transformation. At this stage they select an area to scan in at Max Detail.

firm to operate a 3047 Iris printer.

Through its use of high-end technology equipment developed by Scitex and Iris graphics, The Color Space offers an obtainable environmentally friendly solution for large-format color, duotone, or black and white prints. These prints can be done on fine-art papers and other surfaces for exhibition, limited edition portfolios, as well as for advertising-related comps, and murals up to 47". There are virtually no pollutants used in the entire printing process, which is being developed for archival permanence. In addition to desktop services, The Color Space also has traditional clients in the areas of digital front-end prepress, high-resolution Scitex and Hell scanning, digital retouching color separation, and press proofing.

The Color Space prints on any flexible medium that can wrap around the 47" Iris drum, although some materials may not be porous enough to accept ink. In addition to specially treated gloss, semi-matte, ultra-matte, acetate, and translite stocks optimized for the vegetable-based inks, they also stock high-end archival fine art watercolor and print making papers including Rives BFK and Arches. Clients occasionally provide their own media for projects.

To get familiar with the digital process

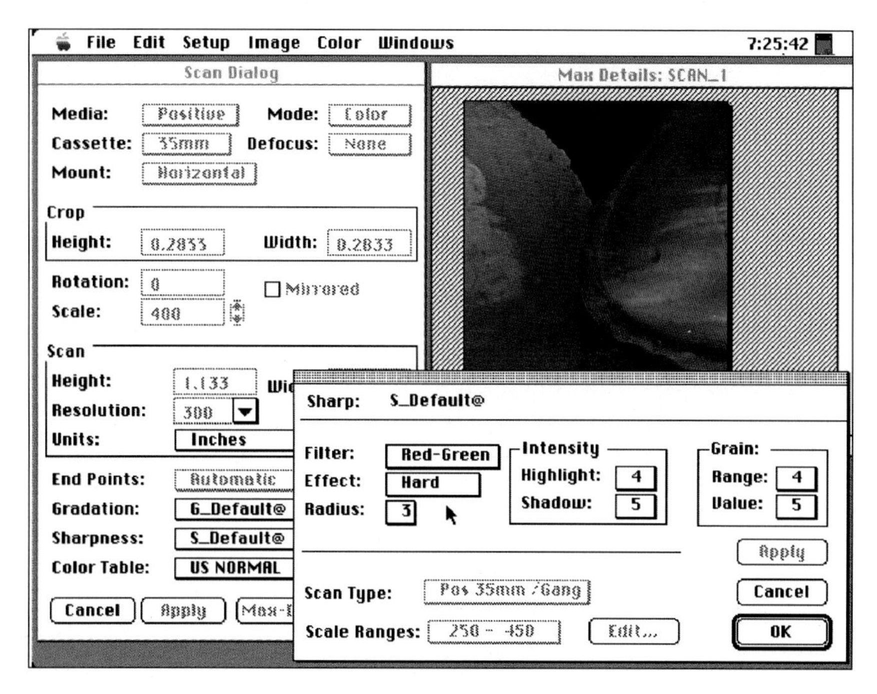

Next they Unsharp Masking, *making adjustments to the amount and size of the sharp effect. The* Intensity *values determine how light or dark the transition of the sharp edge is, and the* Grain *values determine what the scanner sees as grain (and therefore should not sharpen.)*

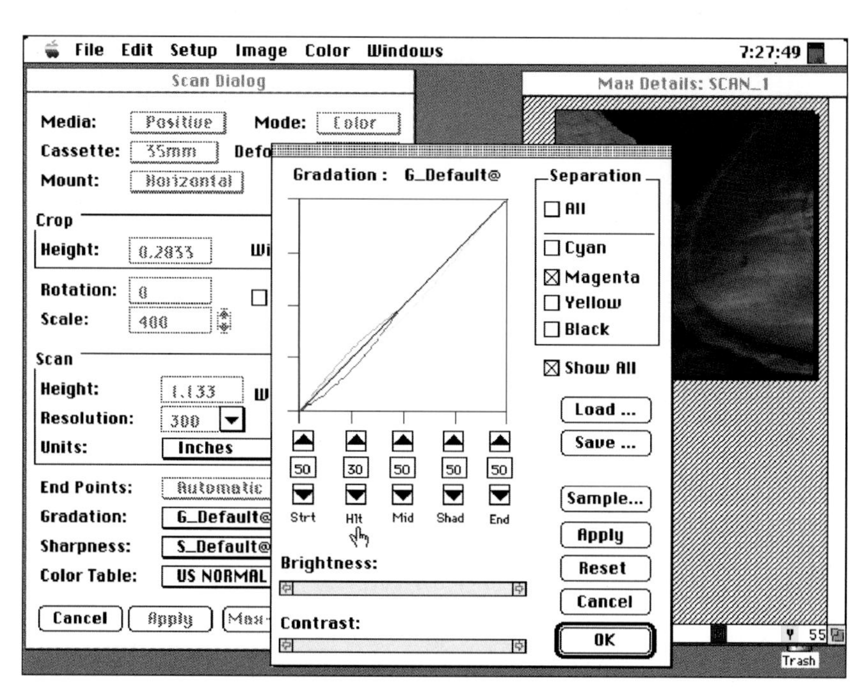

The Color Space can make changes to the "curves" in Gradation. *Other areas used to manipulate color are* Color Control *and* Color Table Edit, *both of which work on the 48-bit file.*

used at The Color Space, let's follow Jody Dole and Peter X through the creation of a high-resolution scan and large-format Iris print.

ELEMENTS OF THE PROCESS

Scanning

In scanning artwork, The Color Space prefers to work with first-generation negative or positive art–film that always produces the best results. Although they work with a Hell 3000 drum scanner (required for oversized art), its Scitex SmartScan is useful in working with clients, since it provides immediate visual feedback when making adjustments to sharpness, color balance, white and dark points, and so on. The SmartScan produces direct CMYK scans using RISC-based algorithmic transformations. The image is transformed on the scanner's "color computer" while still in the 48-bit mode. Not unlike over sampling in a sound studio, transforming the 48-bit file before it is scanned in a 24-bit desktop image produces fantastic results.

In a typical working session they produce a small scan (6Mb) that is printed immediately on the Iris. Upon inspection of the results, the art is rescanned with further adjustments. The entire process for each iteration takes only 10 to 15 minutes, so that in an hour, the artist can view up to five variations printed on the actual media, which

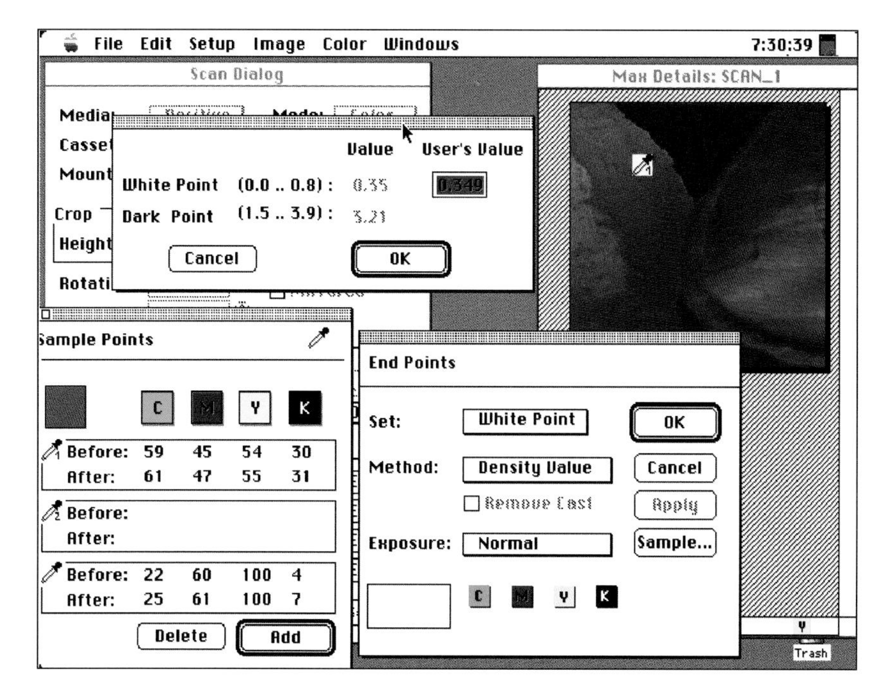

With these settings, they can greatly enhance the ability to capture either highlight or shadow detail, as well as remove any undesirable cast to the image. Again, this is manipulating the 48-bit image.

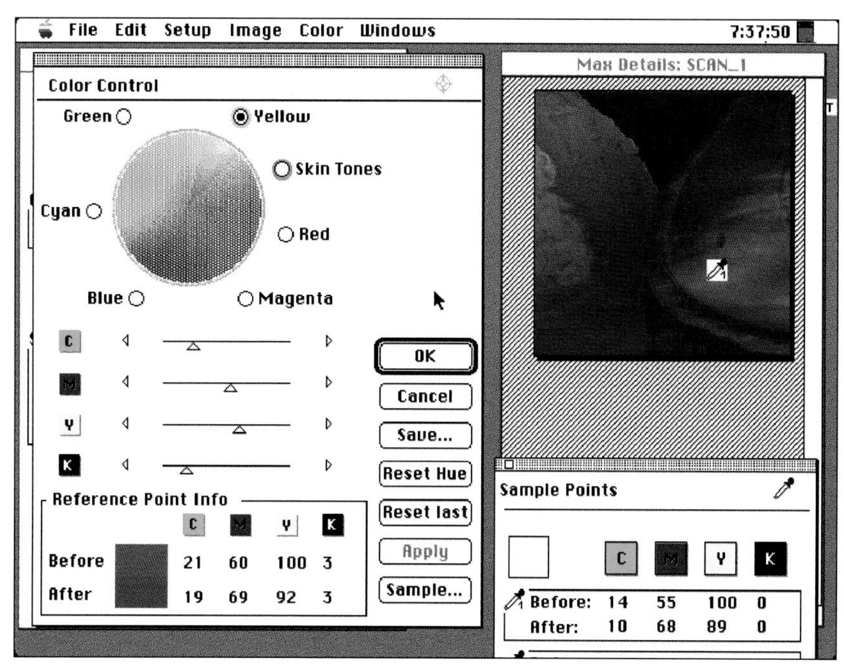

This feature performs edits to the Color Tables, *changing all color in a given hue. Another function called* Color Table Edit *allows them to change specific color. Both affect the image while it is still a 48-bit file.*

entirely removes any guesswork as to what is being approved. Aside from the obvious differences between viewing an RGB image and viewing a CMYK proof, the Iris printer's ability to print on nonstandard stocks (each with its own color characteristic), makes any other method of proofing seem obsolete.

Whenever possible, all adjustments to the image are made at the pre-scanning stage rather than in Photoshop. Once the image is scanned, that data is limited to being recreated through interpolation only (the application's "guessing" as opposed to getting the "raw" data from the image). For example, if shadow detail is not captured at the scan stage, subsequent manipulation can only create the appearance of greater detail.

Printing on the Iris 3047

The Iris is a continuous tone ink jet printer. Physically, it prints at a maximum 300 dpi, but because it is continuous tone, the apparent resolution is 1,800 dpi.

Media is wrapped around the drum. As the printer starts, the drum begins to spin. The rate (100 to 240 revolutions per minute) determines the maximum density of ink that can be deposited by the printer a slower rate allows greater density. The print head (4 CMYK ink jets) begins to travel across

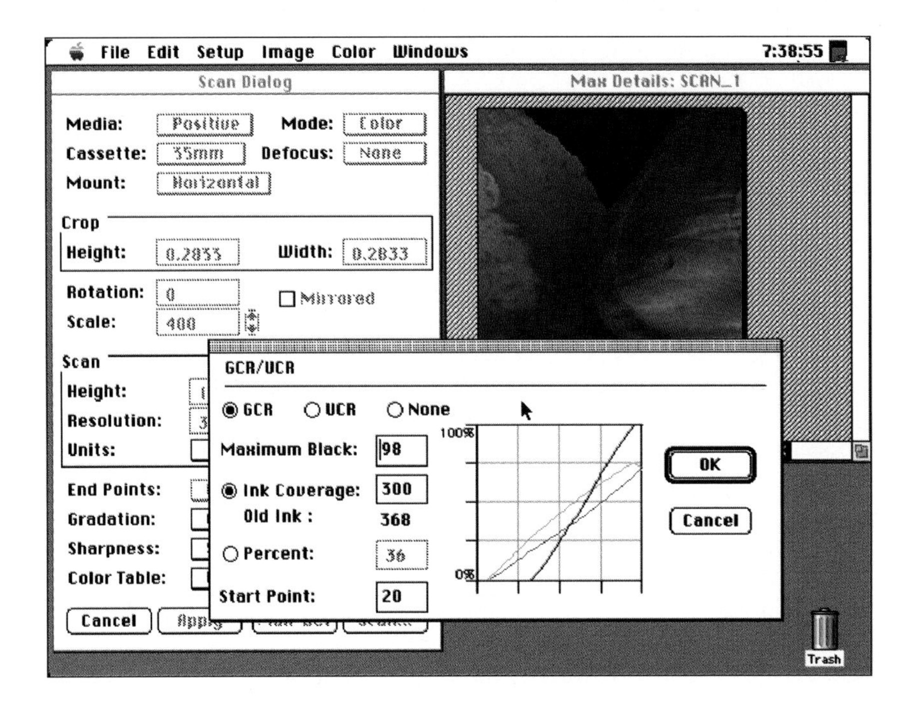

Gray Component Replacement *and* Under Color Removal *values determine how an image is separated for offset and web process printing.* GCR *affects the gray component of all colors, whereas* UCR *affects grays in neutral areas only. Total* Area Ink Coverage *determines maximum values of CMYK after* GCR/UCR. *Whereas these features address real world requirements of typical process printing, Iris printing is not limited to these parameters. This provides greater control over how images are scanned and printed.*

Gray Control *and* Input Gray Levels *determine how gray values are reproduced and transformed into CMYK. Since the Iris has unique characteristics in how it renders gray values, this is an important feature.*

the length of the drum. For a full size 47" print at a standard drum speed of 140 RPM and 300 dpi, the print time is one hour. Each ink jet produces a stream of charged ink droplets. If you take a garden hose and shake it, you make a stream of big droplets of water. In a similar fashion, the ink jets each have a crystal attached to the end of a glass tube which vibrates at approximately 1 MHz. This produces one million droplets per second. Just before the drop leaves the glass tube, it is charged with electricity.

This charged drop now flies through the air until it reaches what is called a *deflection structure*, which is essentially a magnet with a set of plastic knife edges. If the magnet is off, the drop goes over the knife edge and hits the drum. If the magnet is on, the droplet is deflected down, hits the knife edge, and is then vacuumed into a waste bottle.

On many occasions, Jody relies on a high-end desktop system for proofing experimental ideas with the Iris Smartjet 4012 before creating larger 30" x 40" exhibition prints on the Iris 3047.

Using ultra-matte Iris paper and ID inks, the 11" x 17" prints usually end up in Jody's portfolios. His clients and agents agree that the quality of The Color Space/Iris 4012 prints look better than the dye transfers of an era before. They're much

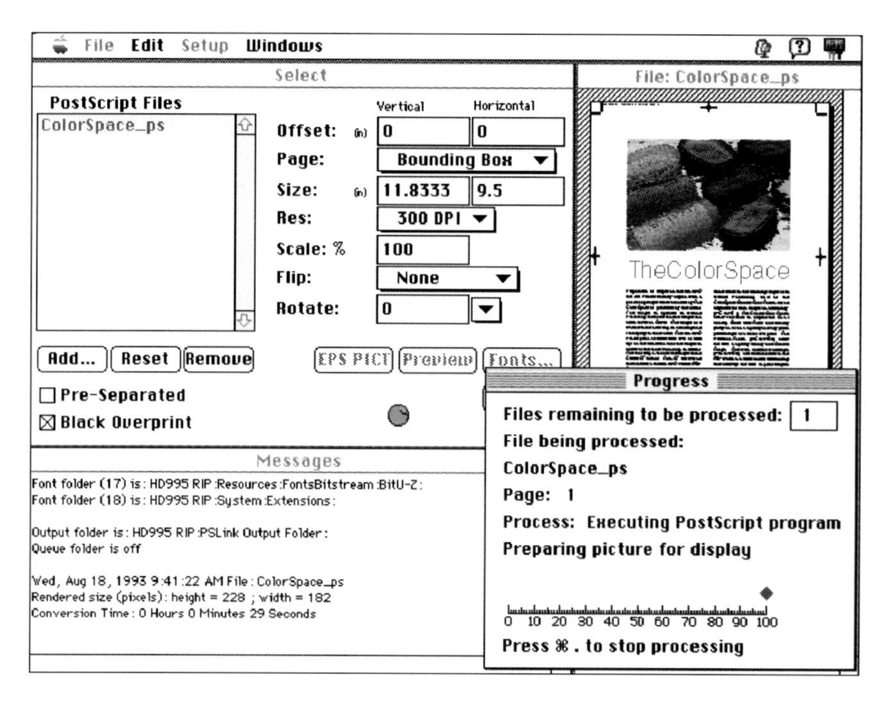

When combining text and images, it is necessary to RIP (rasterize) the composite file. Shown is PSLink, the Mac-based program for creating these files, which are essentially big Scitex CT files. For a full-size, full-resolution image, the file size is 540 Mb. The same program allows scanned images saved in Scitex CT format to be loaded onto the FEP (Front End Processor), which is the DOS-based workstation that controls the Iris 3047.

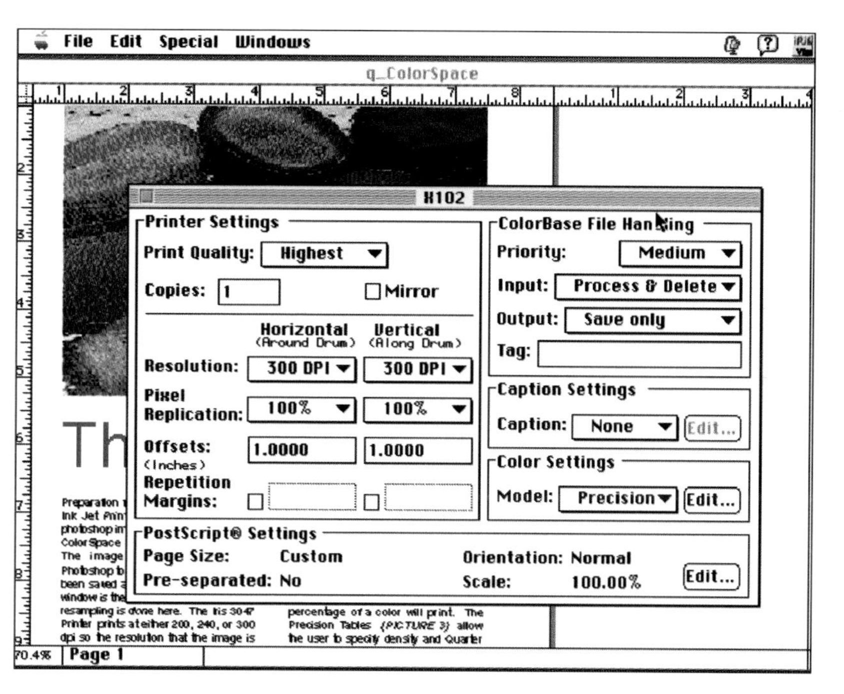

The following parameters are established here: the resolution (the printer has three native resolutions: 200, 240, and 300), the offsets of the print, whether the print gets repeated down and/or across for multiple images, and other file handling characteristics such as priority and orientation.

safer for the environment, too.

In addition to an Iris 4012 printer, the workstation in his photo studio consists of a DSP Accelerated Mac Quadra 950, 20" RGB Monitor, two 1.2 gigabyte hard drives, and an assortment of transportation devices, including DAT, SyQuest and optical disks, plus a Nikon LS3510AF 35mm film scanner. Photoshop is used much of the time. All other scanning is performed on the Scitex SmartScan under the watchful eye of Peter X. Dole comments:

> *When it comes to museum quality printing, no one today has even come close to what Peter X has achieved with this process. He is the digital Michelangelo of fine art printers. Anyone in the world who knows Iris knows of Peter's reputation as a master of the medium. He has a rare talent and has taken fine art Iris printing to another level.*

Jody uses many of the electronic capabilities available in the digital environment in his everyday image making process. He is careful to point out, however, that there is no real evidence of digital imaging in his photography work.

> *I basically use many of the*

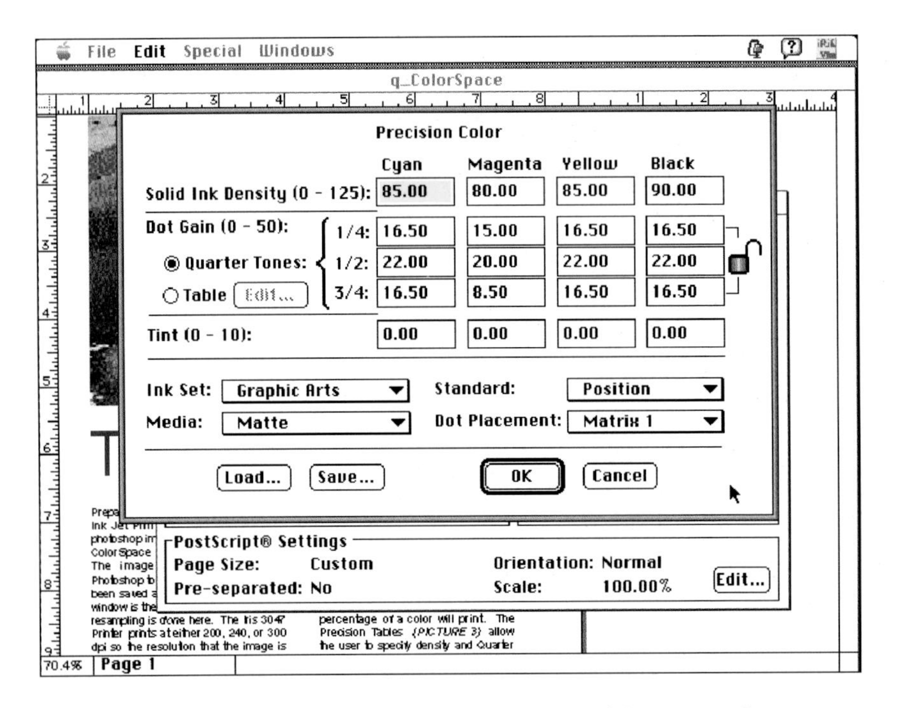

Precision Color was designed to facilitate the use of the printer for prepress proofing. Precision Tables create basic printing parameters, leaving only essential controls for adjusting color. These controls are familiar to the process printing environment and allowing for over-all tints to match specific stocks.

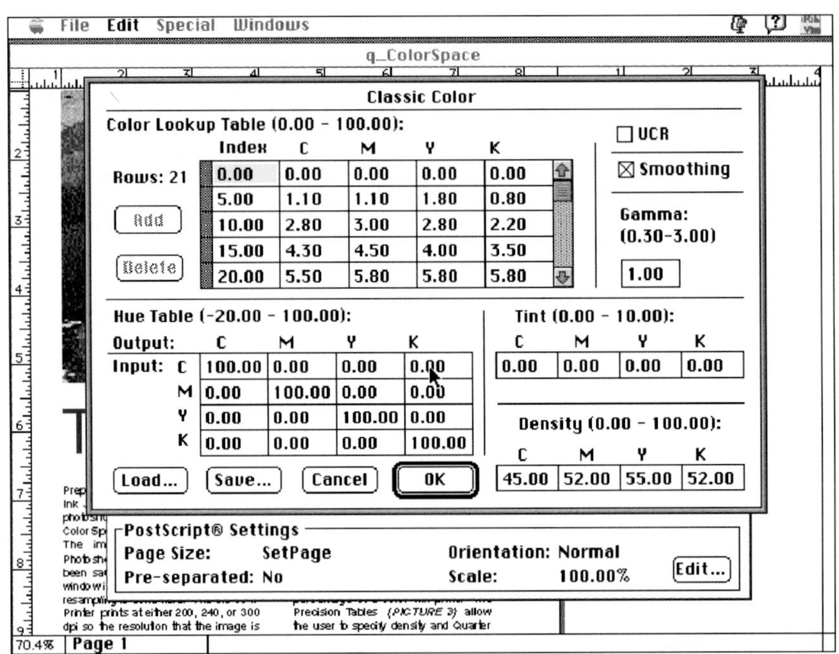

Classic Color is the original program and provides the greatest degree of control. 21-step Color Lookup Tables *are created to optimize the color space of different media.* Hue Tables *change how the basic process colors render. One of the unique abilities of the Iris printer is to multi-strike an image provide saturated images on Translite, the material for backlit images.*

Photoshop tools to enhance what I've already created with my camera. No computer tricks necessary. After thousands of tests, I'm able to get a very accurate beginning right at the scan level, even with the Nikon 3510, by carefully adjusting curves for exposure, contrast, brightness, color balance as well as white and black points. I do this mostly by eye and with the occasional help of computer assisted densitometer. For our higher res work, Peter's expertise on the Scitex smart-scan is a technological mind blower.

Many times though, Jody opts to experiment with his 12-bit Nikon 3510 and Photoshop.

Working in Photshop

After the image is scanned and in the system Jody works in Photoshop and other color programs to fine tune even further by cleaning up dust or scratches, making additional color corrections, sharpening when necessary, selecting and adjusting

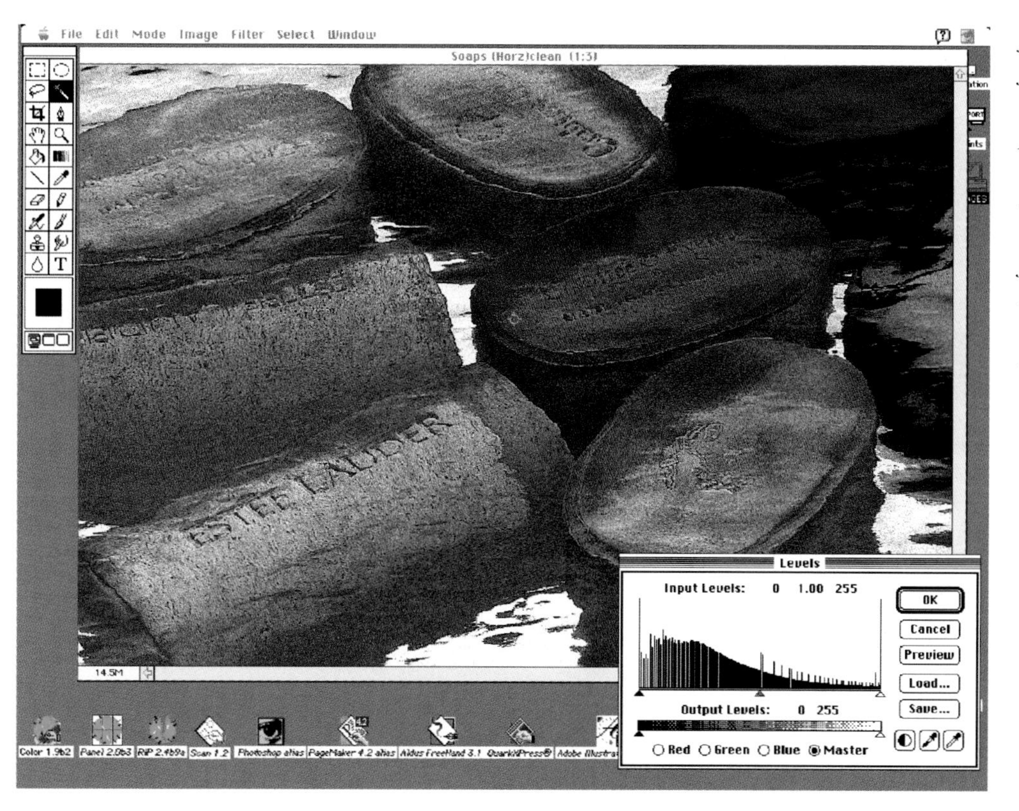

Jody created this image for editorial use in W *magazine. The original photograph was taken in his New York studio, scanned and imported into Adobe Photoshop for additional correction. Here, detail is being opened up in the pre-selected shadow areas using the Photoshop levels control.*

shadow and highlight areas, and from time to time, making subtle adjustments in composition.

Using digital technology, it's simple to remove an imperfection from an object or fix an inadvertent "ripple" in the photographic process. Final print, chrome, or separation output is of higher quality than the original.

Jody notes that when it comes to exhibition printing, there is a lot of information on the calibrated screen that untrained eyes could easily miss. In fact, most times, human error is the cause of undesirable results. For example, as little as a 1% color difference on the RGB screen could yield a 5% change in the output, sometimes more when converting RGB files to CMYK. If an operator is off by as little as 25% the difference by the time it comes off the Iris, if it's off on screen by 10% means the print could be off by 50%. Says Jody:

> *Half isn't even close to being an acceptable margin of error. When it comes to our output, we must be dead on 1000% correct or it doesn't pass. Occasionally, I'll look at a print and say it's too yellow and*

The Iris 3047 printer is shown here with the cover in the open position.

think didn't see it on the screen, then I'll look at the monitor very closely, maybe go in with the densitometer and sure enough, the yellow's there, I just wasn't exact enough and didn't see it. Being right, providing of course that one is calibrated exactly to the output device. At The Color Space, we make it a point to achieve maximum screen/output calibration. Our expertise saves time, money and headache for our clients who work closely with us.

After final adjustments and corrections are made, the image is proofed on the Iris. Smart Jet Tools, a set of two printing utilities, allow direct output of continuous tone images from the Quadra 950 to the Iris 4012 SmartJet printer. Both utilities feature Iris precision color correction, allowing color correct files to simulate SWOP standards, if desired.

In the hands of Peter X and Jody Dole, this process produces prints that are works of art. Jody explains:

In this detailed view, the CMYK colors are being applied by the ink jets.

What we are doing here is adapting high technology available previously to only a select few, and creating the finest quality archival prints known today. We are making the process available to photographers, artists, designers, and corporations who do not wish to invest time in the difficult learning curve or the seven figures it takes for equipment and staff.

EQUIPMENT

- **DSP accelerated Macintosh Quadra 950 and 840AV computers with 128MB ram, 1.2 gi hard drives and 20" Supermac Monitors.**
- **Ethernet**
- **iris 3047 Printers**
- **Iris 4012 Printers**
- **Scitex Smartscan**
- **Hell 3000 drum scan**
- **Nikon LS3510AF Scanners**
- **Dolev image setter**
- **Optical, dat and syquest drives**
- **Adobe Photoshop, Quark Xpres, Iris Smart Jet Tools, Smart Vip, PS Link, Visa**

CONVERSATIONS

JODY DOLE
Partner, The Color Space

Jody Dole who graduated from Pratt Institute gained overnight commercial success in 1989 with his worldwide advertising campaign via McCann Erickson New York. Jody has since earned international recognition for his still life and location photography including top awards from Advertising Photographers of America (APA), The Art Directors Club of New York, Communication Arts Magazine, The Blackbook, Nikon Inc., Art Direction Magazine, American Photo as well as the Graphis photo and design annuals.

There is no real evidence of digital imaging in any of my work and the computer has not at all altered the way I create my photographs, only the process by which I produce the film and prints. All a photographer needs to make the digital process work to his or her advantage is good judgment. When you have the power to adjust the genetics of photography dot by dot, you're in a very powerful place, if you misuse it, you end up creating some pretty weird stuff. I don't intend to split the Flatiron building in half with bolts of digital lightning. I enjoy the architecture just as it is, just as Steichen saw it in 1900.

Peter and I have fine tuned the digital print making process with experience gained from years of involvement with the visual arts and tens of thousands of tests. There were no rules, no guidelines . . . we invented many along the way.

The technology available at The Color Space can combine thought and visual imagery of the very highest level. With a great deal of thought, we are adapting a technology that goes light years beyond conventional photo possibility. That there is no pollution factor in the process is a big plus for us all.

We're constantly exploring new ways of seeing, of showing, of presenting. At The Color Space, we specialize in high end exhibition and limited edition work for clients who really want to go beyond perfect, up to the next level.

PETER X. KSIEZOPOLSKI
Creative Director
Peter X(+C) LTD Designers

The key to our experience is that we are graphic designers who got involved with computer production. Because we bring that sensitivity to our work, and because we intimately understand the workflow from the point of view of the designer and artist, we feel that helped us in our success in working with people.

Seven years ago, when the desktop revolution started, it coincided with our forming X(+C)... previously we all worked together as the Graphic Design department of Pei Cobb Freed & Partners, Architects. We bought a Mac and intended to put it into the computer room. We were rather naive about how we were going to use computers. Since I was the one learning how to use the earliest PostScript capable programs, QuarkXPress and Illustrator, the Mac ended up in my office. It wasn't long before I got tired of kicking everyone off of my desk, and realized that the Mac was, after all, a personal tool. All this was before there were any high-resolution input/output devices. Programs were rudimentary and buggy, so that back then it really was a leap of faith that it would all catch up, which it obviously did.

Part of our early explorations of the technologies included Iris continuous-tone printing. The printer offered what seemed to be the ultimate color output—though at the time the one model could only do 23" x 23" prints and there was no PostScript RIP available. Because our focus at the time was architectural presentations, the larger format and the ability to print CAD files was important. We continued our involvement with Iris and about three to four years ago a PC-based RIP was used until a year later when Scitex and Iris developed a true Adobe PostScript RIP. In an industry where software and hardware life is measured in months, it is amazing how Iris technology has remained essentially the same, a tribute to a great technical breakthrough.

BOB COLE
Creative Director, McCann Erickson Worldwide Advertising

I first met Jody Dole in the summer of 1988. He arrived at my nineteenth floor office early one morning, and presented some work of outstanding quality. Experiments with pottery, bones, metal, glass objects, and simple geometric shapes using very unusual lighting and backgrounds. I saw that he had brought with him a new vision, the talent was obvious.

Jody photographs ordinary objects so that they transcend themselves with incredible beauty. His images emit a striking ethereal light that remind me of classical painting when the artists looked at peaches or olives and made them absolutely enlightening beyond their ordinariness. His way of seeing has a fresh quality that I knew would lend itself really well to any contemporary advertising and certainly to a campaign that I was wrestling with at the time. To Jody Dole, a photograph is not just a record of the way something looks. Rather, it is the coming together of the elements you see in great art: color, mystery, light, shadow, texture, and perspective, all in focus to create emotion. He simply takes photography to another level. Jody sees things for what they are and then shows you what they can become.

Through our collaboration was born an award-winning advertising campaign. My ideas and his photography saw eye-to-eye. Art met commerce and both won.

SUMMARY

The changing digital technology as used by Jody Dole and The Color Space is profoundly changing the face of the photography and advertising industries. Electronic imaging saves steps and money by delivering digital images to an agency or client, reducing drastically the number of steps and vendors it takes to produce a billboard, bus-side ad, poster, or trade show display. The process used to require four studios to:

1. Shoot the photograph,
2. Process the photograph,
3. Add special effects, and
4. Produce output.

Limited edition prints—one or three or fifty—that used to be available on long schedules and at exhorbitant rates, are now easily attainable from one studio with sophisticated results in minutes or days, at a fraction of the traditional cost.

AD SLICKS

This chapter takes a look at the world of the advertising designer, who often requires a breadth of knowledge that extends beyond design to marketing, strategy, and print production. Learn how Media Designs executes an advertising campaign for the Associated Press.

MEDIA DESIGNS/THE ASSOCIATED PRESS
Mobil 1
336

The Associated Press (AP) has a series of news services it delivers to radio stations, as well as to television stations and newspapers. News services such as this are referred to as wire services. These wire services formerly delivered their news via a printer. With the coming of the computer age, delivery now comes in the form of a dedicated computer system and print station.

INTRODUCTION

The Associated Press (AP) recently revamped a series of services that go over their news wire, which they wanted to promote to radio stations.

The service provides news, actualities, reports, and so on. Until they reworked their system, AP had a printer at subscribing radio stations, for which subscribers had to sign a five-year contract. The AP service was expensive, and in the radio business, which is very volatile, a five-year commitment is a big deal.

These printers generate stacks of pages of news each day. The receiving radio station then had to review this extensive material and decide what would go on the air. AP's old service was challenged by lower-priced, fax-transmitted services that presented highly condensed material. This condensed material was attractive to radio stations because they didn't have to sort through and rewrite the stories.

AP contacted Media Designs to do the advertising for their new information delivery service. AP had two goals for their advertising: to announce their new information services, and to convince people that it was worth making a five-year commitment at a much higher price than the fax services. The focus for these concepts was the image of the Associated Press as a well-established institution,

One of the major elements on the cube had to represent the sports service available on the AP wire. Scott Randall selected an image from the Olympics, since it is such a timeless event.

and the quality of the material that would be delivered by the service.

ELEMENTS OF THE PROCESS

It was very important to the Associated Press not simply to introduce the new look of what they were sending to the radio stations, but also to give a real prestige image to it. From the beginning, it was stressed that whatever Media Designs did had to be graphically excellent, which was the deciding factor in going with four-color ads.

Selecting the Images

The AP product had four different package options: a show prep package, which contains one-liners for morning show Disc Jockeys; a soft news package, which is entertainment news; a sports news package; and a hard news package. All of these packages had to be integrated into the composite ad. The composite ad was running in four different publications with different requirements. Media Designs had to prepare ad slicks for each publication, but each ad had to have the same look.

Media Designs looked for photographic images that were timely yet evergreen–images that were with the times, but that wouldn't quickly fade. For example, the elections were approaching, but they didn't want to go with images of Perot, Clinton, or Bush because only the winner would stay in the news.

They chose four main images. The Olympics were being highly promoted at the time, so they used the long-jumper to bring across the sports idea. Boris Yeltsin and the Los Angeles riots were fresh in the news. They felt Boris Yeltsin would be around for a while, and the Los Angeles riots continue to be a source of discussion. Madonna, says Scott, never goes away, so that picture was chosen to represent entertainment news.

For the side images they selected the flood shots, since weather is an integral part of AP's hard news service. The space shuttle is also an evergreen hard news item. The Bosnia-Serbian war had enough of an impact to be around for a long time,

An image of Boris Yeltzin was selected to reflect the hard news available on the service. While the elections were pending at the time of the campaign, Media Designs decided to go with Boris Yeltzin because they thought he would be around long after the elections. They were right—he was still in the news a year later.

and the Disney image was chosen to support the soft news concept.

Once they found images that provided a nice contrast to each other, Media Designs mapped the photographs on every cube until they found the right combination, visually and ideologically. Scott Randall, President and Creative Director of Media Designs, explains some of the thought that went into image placement:

> *We felt that Madonna, because she had the black background , made a good bottom one because it held up; it gave us a border along the back which was more defined than some of the other ones.*
>
> *We liked the top with Yeltsin because it had all that red, this was a rather busy image so it —the red—gave us a border there. These images could easily have run into each other, and that's where some of the retouching got involved where we went into the edges and either darkened up or lightened up things so that they looked better against the image above or below them.*

The Associated Press has a vast photo library in Rockefeller Plaza, a few blocks from Media Designs' offices. Scott did the photo editing himself, looking through hundreds of images to find the ones that would portray all of AP's services.

Media Designs began producing the ads and went through the copy-approval process. When it came time to create the artwork, they ordered slides from AP's photo labs, then sent the slides to a service bureau to have high-resolution drum scans done. Media Designs then retouched the photos for color and contrast in Adobe Photoshop.

The photo files were huge. When it was time to render the three-dimensional cubes, Media Designs had to strip down its entire hard drive of all of its software and data files except for Autodesk, which was required to render the cubes.

Rendering Three-Dimensional Cubes

The original concept for the ad campaign was to use photographs. It was important to convey the newness and the cutting-edge aspect of the product they were delivering, so they needed to have a visual image that was not just average. They wanted to do something that really reflected the times.

The client suggested using a filmstrip down the side of the ad. Media Designs developed the idea of mapping photographs on objects of some kind, to give an image-processed look to the ad. The final concept was a cube composite, which the client liked immediately.

A scene from the Los Angeles riots was chosen to further convey the concept of hard news. Due to its powerful and provocative nature, it was a good complement to the Yeltzin image, and is still a topic of discussion.

Media Designs prepared laser proofs created using the 3-D module in CorelDRAW 3.0. However, the module created more of a 3-D simulation than a totally realistic rendering. At that point, they realized they had to go to a full 3-D program to execute the ad properly and decided to use Autodesk to render the three-dimensional cubes.

Using Autodesk was a relatively simple process. Alkis Papadapoulous, Media Designs' technical consultant, created a cube, cloned it four times, placed the photos on all four sides, stacked them, then rotated them.

Using a three-dimensional rendering application is different from using a two-dimensional one. Aside from the additional dimension, 3-D programs provide many lighting options and camera angles. Explains Scott:

> *You're getting away from your traditional flat 2-D print idea. What you're really doing is setting this up as though it were a photo shoot, and you're lighting it as though you had built these cubes, pasted photographs on them, and were now taking the photograph. You had to light it. There were shadows.*
>
> *Shadows are another thing we should talk about. What we ended up doing on these was creating a cube, cloning it four times, stacking them, and then rotating each cube to try to maximize the view that you're going to get. In other words, some things are sort of straight-on, others are side viewing. We chose the four front ones as the main images, and the rest of them were support images. That was the concept behind the photograph placement scheme. We basically just rotated them until we got just the right amount of space to give it that 3-D look.*

Media Designs experimented with putting colors on the tops of the cubes, but it started to "fight" with the photographs. They decided to use the gray/black scheme for the tops and bottoms of the cubes, and to light them in two ways. Autodesk allows you to self-illuminate an object, which means that it glows by itself. Then you can apply outside lighting sources. This has the effect of setting up a photo shoot scenario in the software program.

The Autodesk rendering created some very large files. The images were output at 2,000 dots per inch (dpi), which is high. Their computer was running with 32Mb of RAM, yet even the smallest retouching job went slowly. They had to walk the fine line between keeping the files man-

The image of Madonna reflects the entertainment news available on the wire.

ageable while making the resolution as high as necessary to avoid an aliased look. Media Designs ran a few tests before they were satisfied that the image they were getting was really anti-aliased and solid.

Finalizing the Ads

The page was constructed in Aldus PageMaker 4.2, and type was set using a typeface called Matrix. Media Designs created a low-resolution file for the cube illustration, and placed it in the Aldus PageMaker file for position.

Media Designs output the 3-D rendering as a TARGA file (see the sidebar on the TARGA format),which was then output by a service bureau to an 8" x 10" chrome transparency. Before the final output, a few last minute touch-ups had to be made. Since the original rendering was done on a PC and the service bureau worked on Macs using Adobe Photoshop, this was a cross-platform event.

For the most part, this process went smoothly, except for one of the shadows, which came out very differently from the one that was created in Autodesk. Somewhere in the transfer, the fall-off on the shadow changed to a hard edge, and nobody could figure out why. The service bureau then had to manually airbrush the correct shadow directly onto the chrome.

They then output the image on resin-coated (RC) paper using a high-resolution image-setter, and had the 3-D rendering separated on a Scitex by a service bureau. Each of the four versions of the ad had to be output separately, since they all required different line screens, from 85 to 133. The ads also varied in size, from 5.5" x 6.5" to 11" x 13," and one was run as a black-and-white duotone.

Maintaining Image Integrity

With the ad that was to be run at a line screen of 85, Media Designs had to be concerned with the issue of dot gain. Dot gain occurs when a print is made on newsprint and the image spreads due to the porous nature of the paper. There are a lot of color density requirements that are fairly strict. When a print is made on a coated stock, like the magazine format of *Broadcasting* magazine at 133 line screen; dot gain is not an issue, since the paper can easily hold the image. On newsprint however, the color spreads and spots can appear, so images often need to be manipulated further.

Another concern is with images that are enlarged to fit a tabloid format (11" x 14"). For one of the versions (a tabloid format printed at a line screen of 85 on newsprint) Media Designs had three layers of image-degrading factors to contend with: The cubes were increased about 250% to fit the tabloid format, which degrades the image; it was printed at an 85 line screen, which further degrades an image; and it was printed on newsprint, which further degrades an image. Scott explains:

61st YEAR FIRST IN TELEVISION CABLE RADIO SATELLITE 1992 □ $2.95

❑ *BAYWATCH* BASKS IN LIFE AFTER NETWORK / 15

❑ ELECTION UNLIKELY TO MOVE AND SHAKE KEY HILL INCUMBENTS / 24

❑ TV NETWORK REVENUE FLAT IN SECOND QUARTER / 31

Choose Quality.

AP Associated Press

News • Graphics • Software • Radio Networks

800-821-4747

A CAHNERS PUBLICATION

CLINTON'S ARMY

FIELDING TELCOM'S FIRST TEAM

Vol. 122 No. 34

The stacked cube image was run on the cover of Broadcasting *magazine with the tag line "Choose Quality."*

The Power of Information.

These changing times demand accurate, reliable, and instant news coverage. Delivering that kind of service takes an investment in people and the latest digital satellite technology. In an age where your ability to deliver the news on-time, everytime means the difference between failure and success, it pays to go with the service that has the resources and experience to do the job.

With the announcement of our new two-year contract terms and mornings-only AP DriveTime wire service, the power of the world's largest news-gathering organization is within reach of virtually every radio station. There simply isn't room in this business for anyone who underestimates the power of information. In our 144 years...we never have.

AP Associated Press
The Power of Information.

These two veloxes were provided to different publications as camera-ready art. Directed at different audiences, they use the same image, but different tag lines: The Power of Information and Choose Quality.

When AP hired Media Designs, it had just revamped its wire service, not simply in terms of its technology, but in terms of its product. Designed to reflect the way our culture digests information—short, segmented, predigested messages—AP's new product collated information into a much more manageable form.

The mission of this campaign was to announce these new services and convince subscribers that the service is worth the required five-year commitment. Although the AP service is costlier, it offers a high quality product. Thus, Media Designs had to not only introduce the look and feel of the new product, but to add the appeal and value of a prestige image. The most important factor of the campaign then became graphical excellence.

PREPARING AN AD SLICK

Ad slicks are prepared so they can be sent to a variety of publications. Slicks are usually prepared using veloxes. A velox is a high-resolution, camera-ready form of a stat that can be shot at different line screens. Where stats are normally used to place art for position only, veloxes are camera-ready.

Newspapers and magazines use a standard unit of measurement, called a standard advertising unit (SAU). Ad slicks are prepared to fit this unit of measurement. Using an ad slick prepared in SAUs is a way of having camera-ready art without having to create a mechanical. Since different publications use different line screens, different versions of ad slicks often need to be created. Some agencies will prepare ad kits that contain a variety of ad slicks created to a publication's specifications. An agency can then purchase space and simply call the publication and tell them to use, for example, ad A or ad C for that issue.

These are four-color photographs. It's not like a graphic thing where you have some sort of a processing or an effect where the eye perceives that this is the effect you were looking for. People know what Boris Yeltsin looks like and they're used to a flesh tone that is like video; anytime you have flesh tone involved, there's no room for error. Flesh is flesh and it's got to look that way, and it was difficult to reproduce it in a tabloid format. We worked with the magazine as tightly as we could but we were much happier with the results when it was a 133 line screen coated (paper) situation.

The Targa Format

The TARGA format is something that Media Designs discovered when they started passing images between print and video. On the Associated Press project they needed a format that everyone could read, and a TARGA file is like a TIFF file in that it's a very good platform extension (meaning that the file can be used on different types of systems). Another part of their business is shooting

Photo: Anne Marie Sconberg

EQUIPMENT

- 486-33Mhz IBM-compatible computer with 16Mb of RAM, a 250Mb hard drive and a 1.2G hard drive
- Microtek 600 dpi flatbed color scanner
- SyQuest 88Mb removeable media drive
- 21″ Hitachi Color Monitor
- Hewlett Packard Color DeskJet non-PostScript printer (for proofing)
- GCC-BLPIIS 300dpi PostScript laser printer
- Wacom wireless digital pressure-sensitive mouse and pen tablet
- Iomega 250Mb tape backup system
- CorelDRAW!
- QuarkXPress for Windows
- Fractal Design Painter
- Autodesk 3-D Studio
- Aldus PageMaker
- PhotoStyler
- Aldus Persuasion
- Adobe Type Manager
- Asymetrix Toolbox
- Adobe Illustrator

W·A·S·H 97.1 FM

The Sound of Washington

industrial films and commercials, and they were actually creating graphs and charts in CorelDRAW using its slide-making module (CorelSHOW). They were outputting print-application generated material, and outputting it through a TARGA board and a software program called Video Maker directly to three-quarter inch video tape in house.

Scott says they use it extensively:

> *One of the things that we've always done is create an integrated look for clients, which means that we output "printfiles" to video. In other words, if a client has an annual report that they want to do, and they want to have an accompanying video, I can take frames of video and put them in the annual report. I can take the photographs from the annual report and put them in the video.*

Media Designs is currently working on a campaign for radio station WASH in Washington, D.C. The package consists of a promotional cassette, a subway/bus card, and a commercial. All three of them have the same logo, which was generated in the same place and output to a TARGA file. The promotional pieces have the same color scheme, the same typefaces, and the same gradients. The print logo that is on the cassette card is fully animated in 3-D on the station's commercial, so the public gets a real sense of continuity.

CONVERSATIONS

Photo: Anne Marie Sconberg

SCOTT RANDALL
President and Creative Director, Media Designs

Scott came to New York as a musician, and ended up working in radio as a marketing executive. From there he joined Warner Brothers Music, where he learned print production. He spent three years producing four-color music folios for Warner Brothers while building up a freelance business to make ends meet. When he got to the point where his freelance business was bigger than his job, he decided to leave Warner and form Media Designs, which produces print, video, and multimedia presentations.

You have these kids coming out of school who can design their brains out on the screen, but when it comes to actually outputting film, it's a problem. And it's only been made worse by computers.

We had a job once where one of the young women that was working for me had white type on black paper or white type on brown paper. We had specced out this fancy Italian Filare paper and the client was wild about it, but we ended up having to hand silk-screen the white type on there, and on the black sheet we had to use a drop-out and paint the edges of the paper black.

You really have to think about production. When you work in production everyday, the first thing you look at when you see something is "How am I going to get this onto the printed page? Can it be done?" It gets overlooked a lot, and because I had that print production background, I've always had an eye to "This is a great idea. Can we do it? And if we can't do it, let's move on to something that we can do."

I went into it computers with the idea that I was going to configure a Mac system like everyone else, and PCs never really occurred to me. Windows at the time was a new thing and I just didn't really know that much about it. I have a Mac sitting on my desk so, I was comfortable with the Macintosh already. We're in the weird situation where I do all my correspondence and accounting on a Mac LC30 on my desk, and all our graphics on a PC. It was at that time that I hooked up with Alkis [Papadapoulous] and his background was essentially DOS.

For the money that I spent with a PC system, I ended up with a 24-bit Targa board, a full-color 24-bit scanner, a Wacom digital tablet, a 20-inch monitor, a 486-33 with a 210 [Mb hard drive], a bunch of software, and a Hewlett Packard DeskJet color printer. For the same money, I wouldn't even have come close to buying that kind of a system with a [Mac] fx or anything. The fx has since become, not obsolete, but is not in production anymore at Apple.

continued on next page

All of my clients work on PC's. I have software clients who write software on the PC. I felt at the time that, for the money, I got a very high-end graphics station that I wouldn't have been able to afford on an Apple platform. All my files are interchangeable with my clients' right away, and really the only down side of it, which was substantial for a while there, was that the output bureaus were all using Macintosh equipment. And that only became a problem when, after giving them an EPS file, there was a problem with outputting it.

If it's Tuesday night and the film has to go out Wednesday morning, or has to go out that night and you send your PC-generated EPS file down and there's a problem, you can't get into it to do anything about it in time. It's possible, but it's a huge job and nobody wants to do it so they send it back and say you can't do anything, as opposed to being able to open the native application on the Mac platform and dive right into it. So there were problems in the beginning getting work output. That was the big thing, and that's not to say it couldn't be done.

A couple of things have happened. As you go through dealing with this kind of stuff, the same problems crop up over and over again, so we've learned to sort of go around them to a certain degree. One pleasant development has been that, because of the low price of the PC platform, an increasing number of students and schools are working on it.

You go into [school] labs these days and instead of just Macs, these days, you see PCs running Windows in there. And for a kid getting out of college whose parents want to buy him a graphics work station, the PC is really a much more affordable opportunity.

The schools are doing it, the kids are doing it, and the output bureaus are beginning to realize that there's a market there. One of the output bureaus that we have now has a dedicated PC guy running PC Windows applications. So now when we send them an EPS file, we also send them a native Corel file and, if there's a problem, they're able to go into Corel, make whatever changes have to be done, and still get the job out. So that's been a big improvement.

Up until four or five months ago, finding people who could come in and be comfortable with extensions and directories, as opposed to folders and documents, was a problem. People are intimidated by having to type a file name and a dot and a backslash. It's been a problem.

SUMMARY

More and more advertising designers, because of their increasing breadth of knowledge, are being requested by clients to replace traditional advertising agencies. Computers now give these designers the added flexibility, speed, and capabilities they need to really begin competing with traditional ad agencies. Many in fact, have gained agency status to be able to place for their clients ads with newspapers and magazines. Once again, computers are turning traditionally black-and-white areas gray.

CHAPTER 11

ELECTRONIC ILLUSTRATION

This chapter gives you an overview of the emerging world of digital illustration. Learn how artist Adam Cohen uses Adobe Photoshop to create innovative, vivid, and memorable images.

ADAM COHEN

Spec's Music President Ann Lieff contracted CNI Walker Group to produce a series of murals for the Specs Music chain of stores in Florida. The mural above guides music customers to the country-western section of a store.

INTRODUCTION

Adam Cohen produces electronic illustrations, drawings, and sketches for signage, murals, book covers, and other formats. He is currently working on a job that involves nine murals for Spec's Music, a chain of over 60 record stores in Florida and Puerto Rico.

Adam was retained by Spec's design firm, CNI Walker Group, to create the murals that will be used as visual signage to direct customers to the area of music inside the store: country and western, jazz, rock and roll, pop music, video, classical, easy listening, Broadway music, and Latin music. Spec's may also use the images from the murals for T-shirts and other types of promotion for its stores.

Adam spent a lot of time discussing the kinds of images that should be in each mural with Christine Walker, the project's Creative Director. Once they agreed on the concepts for each mural, Adam just jumped right in.

Adam has developed a process of working in Adobe Photoshop that provides him with the freedom he needs to experiment and work intuitively during his creative process. For him, this places spontaneity first, and the more technical aspects last.

Adam's first step is to create a low-resolution sketch of the mural.

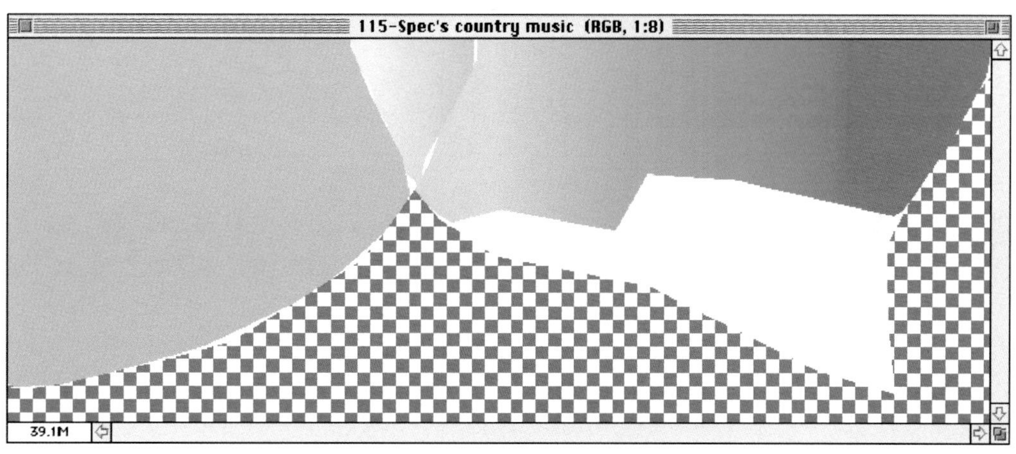

Adam uses the sketch to determine the location of elements for the final, high-resolution art. He typically begins by creating a background.

Beginning a Sketch

Whenever he starts a sketch, Adam opens a window in Photoshop and sizes it according to his client's specifications. For a sketch, he works in RGB color at 150 dpi for faster processing.

Next, he opens other windows and creates objects and patterns that he'll use to compose. His method of composition is to copy and paste these objects until something he likes begins to emerge. Since Adam is working at a low-resolution, the computer performs quickly. This allows him to work out a rough version of what the final art will become during the sketch stage, with shadows, selected colors, and special effects.

When this is complete, he prints the sketch and faxes it to his client for approval. If there are any changes that need to be made, he can easily go back to the sketch and make adjustments.

Creating Final Art

Once Adam has client approval to create final art, he raises the resolution of the sketch from 150 dpi to 300 dpi. The sketch then becomes the mask to create the final art. This ensures that the placement of final art is the same as the sketch. To Adam, creating the final art requires a slower and more technical process, because his focus is to produce sharp, clean edges at high resolution.

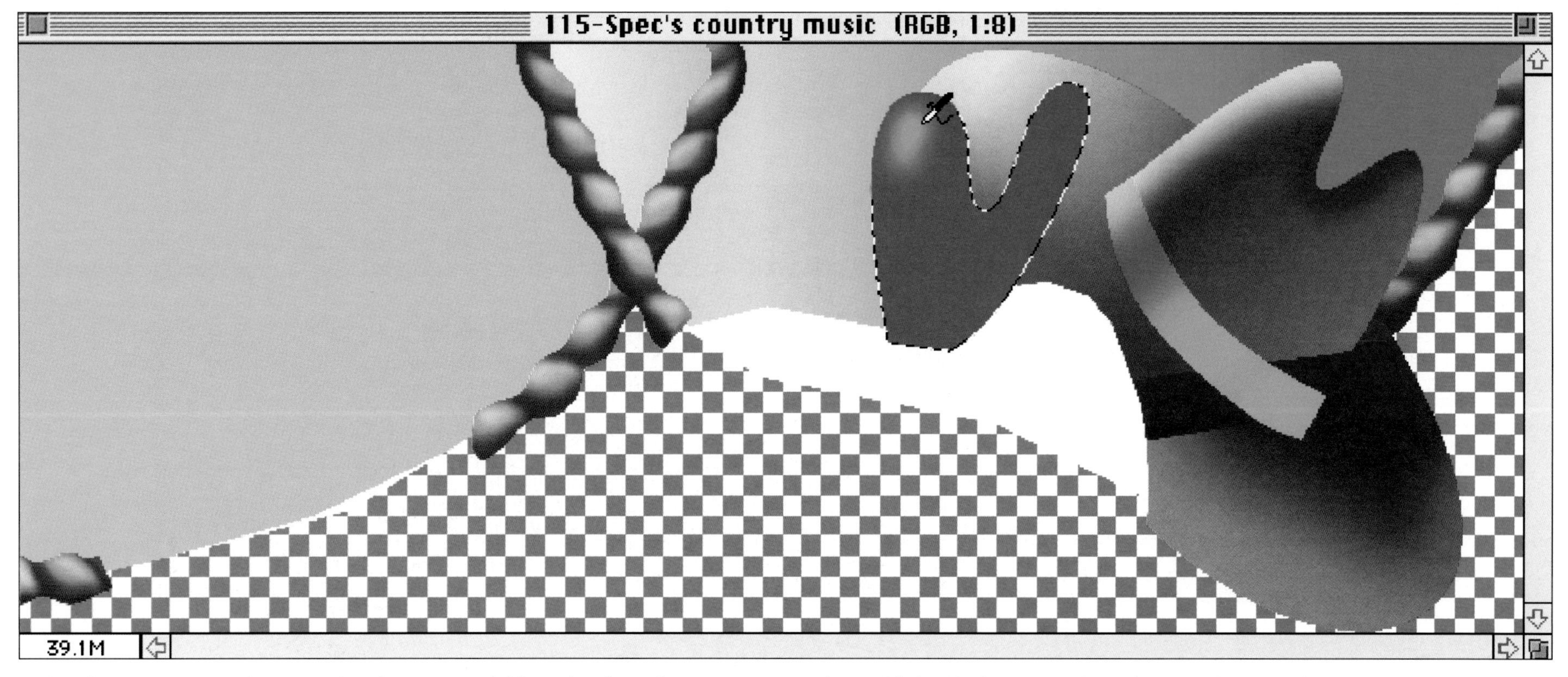

Adam begins to create the cactus by drawing and filling the shape for one piece. He then adds highlights using Photoshop's Airbrush tool.

Adam begins drawing the final art by copying his sketch. He chooses **All** from the Select menu, **Copy** from the Edit menu, and then opens a new file. This procedure creates a new window in exact proportions to the 300 dpi sketch-mask. He then saves this new window with the label "final art." Next, Adam makes selections in the sketch-mask using the Pen tool, and, using the File menu, saves the selection to create an alpha channel.

Demystifying the Alpha Channel

Alpha channels are one of the more difficult concepts for new Photoshop users to grasp. It is also one of the application's most powerful features, and set it apart from the competition when it was introduced.

Until Adobe Photoshop, one of the largest drawbacks for bit-mapped painting and editing programs was their inability to properly maintain the layers artists have come to take for granted in drawing programs such as Adobe Illustrator. Photoshop allows each file to have up to 16 8-bit masks, or alpha channels, that are saved with the image. Since the alpha channels isolate part of an image, they can function as layers, masks, or both. Alpha channels are saved as separate layers, so each one can be processed independently, with any effect available in Photoshop.

Next, Adam opens the alpha channel and selects **Calculate...Duplicate** from the Image

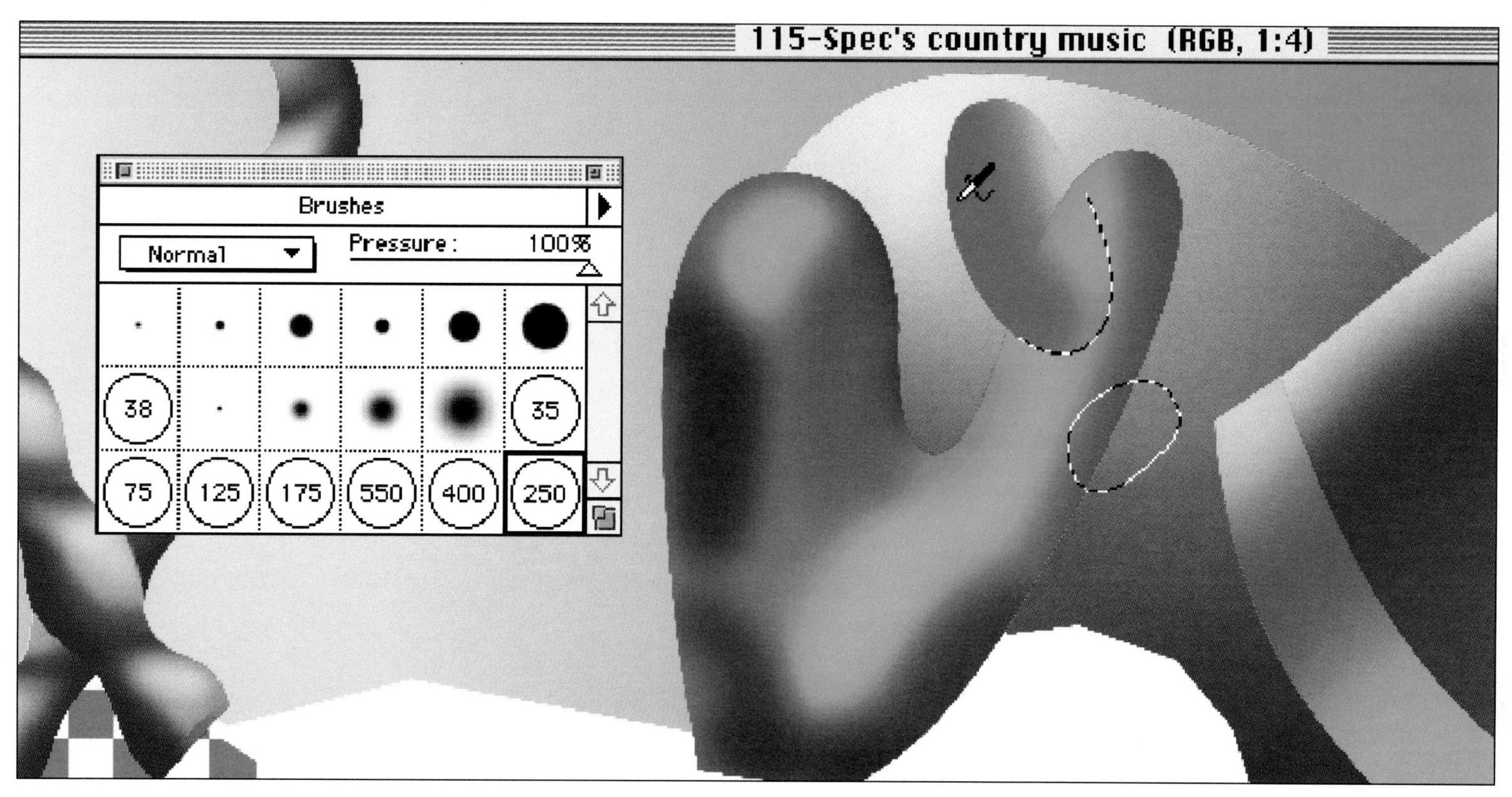

The cactus is further developed by adding additional branches and filling and highlighting them. In this image, Adam uses the floating Brush palette to adjust the Airbrush tool.

menu. This opens the Duplicate dialog box, which allows him to send the selection from his alpha channel to the final art. Photoshop 2.5 has a feature that allows Adam to preview an effect in an alpha channel. If he doesn't like it, he simply clicks off of it, and the image is restored.

Adam repeats this process of selecting one or several parts of his sketch, always trying to start with the largest background area and work as if in layers to the smallest objects. He then sends each selection to the final art window, and fills it with color, blends, and special effects. He uses the Pen tool when selecting areas in the mask, because it has sharp, clean edges. He saves over the same alpha channel each time, unless he knows that he'll need a particular channel later in the painting process. This way, he doesn't have to change the settings unless absolutely necessary.

When the final art is finished, he always views his work at full size (1:1) to inspect it for pixelated edges or any other kind of flaw. Then he takes a closer look at double magnification (2:1).

Using a Color Palette

One of the constraints of the project was for Adam to work with a specific six-color palette that would fit in with the whole record store.

The palette was selected from a selection of Pantone (PMS) colors, but Adam creates each mural in RGB color, then changes the colors to CMYK before generating output. The murals were

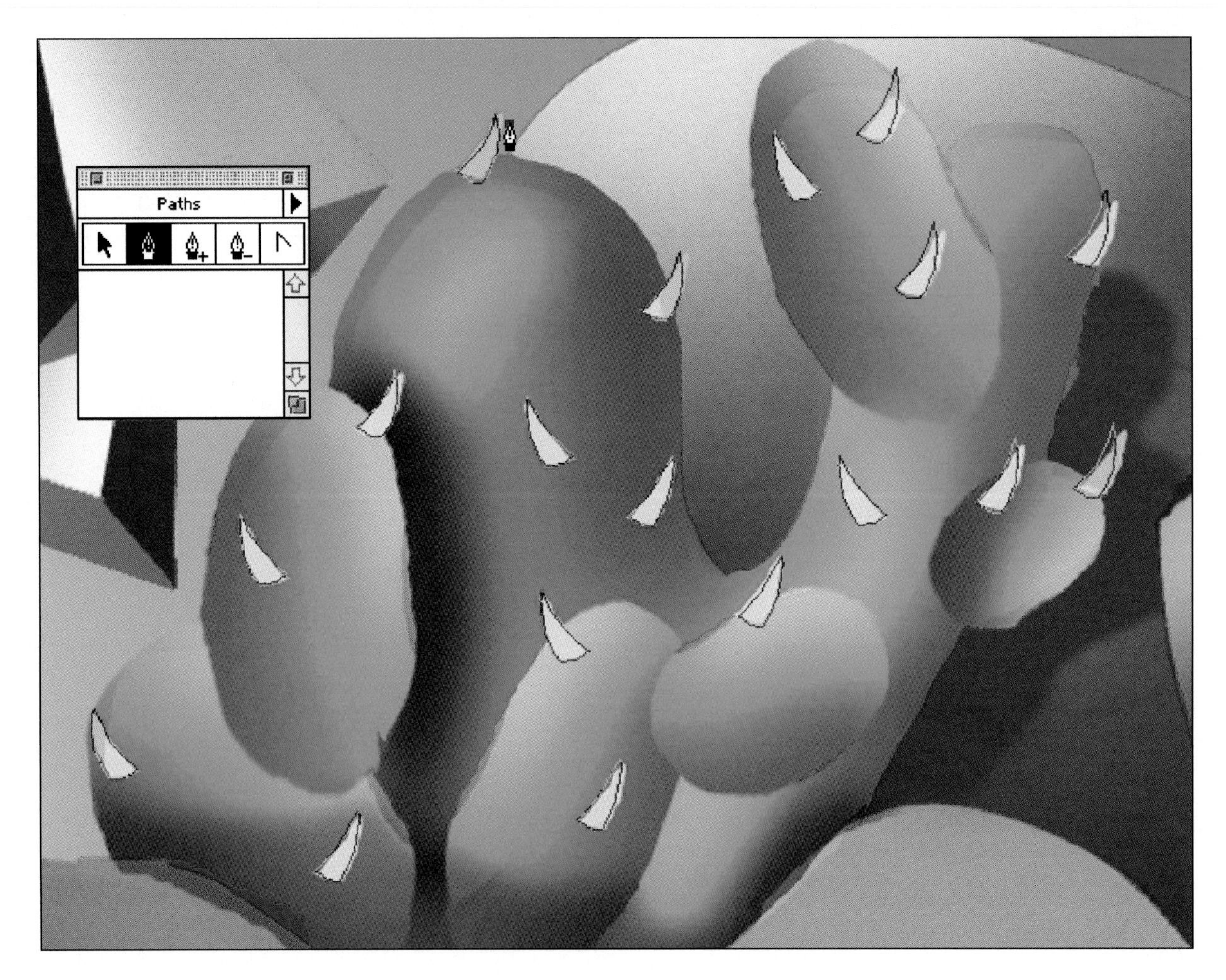

Once all of the arms have been added to the cactus, Adam adds needles to the cactus.

proofed using Iris prints, which are continuous-tone color outputs.

Building the Country-Western Mural

For the country-western mural, Adam first created the lasso rope to break up the composition in a certain way. Then he generated the gradation on the top part of the rope, and the checkered pattern on the bottom. From there, using some reference material for the instruments and stylizing them, he created the banjo and the guitar in separate windows. He created the boot, the hat, and the cow bells next, and played with their positioning, since he had them all in separate pieces. Frequently, he likes to archive images for future reference.

Surprisingly, Adam hardly ever uses his scanner or pen tablet. For all nine murals, he only scanned for tracing one image—sheet music. For drawing, he uses his mouse much more often than his pen tablet, particularly since Photoshop 2.5 introduced its Pen selection tool. He's been using it with the mouse and finds it gives him the sharpest edges. He still uses the tablet when he needs to have a line that looks like it was hand-drawn, or for signatures.

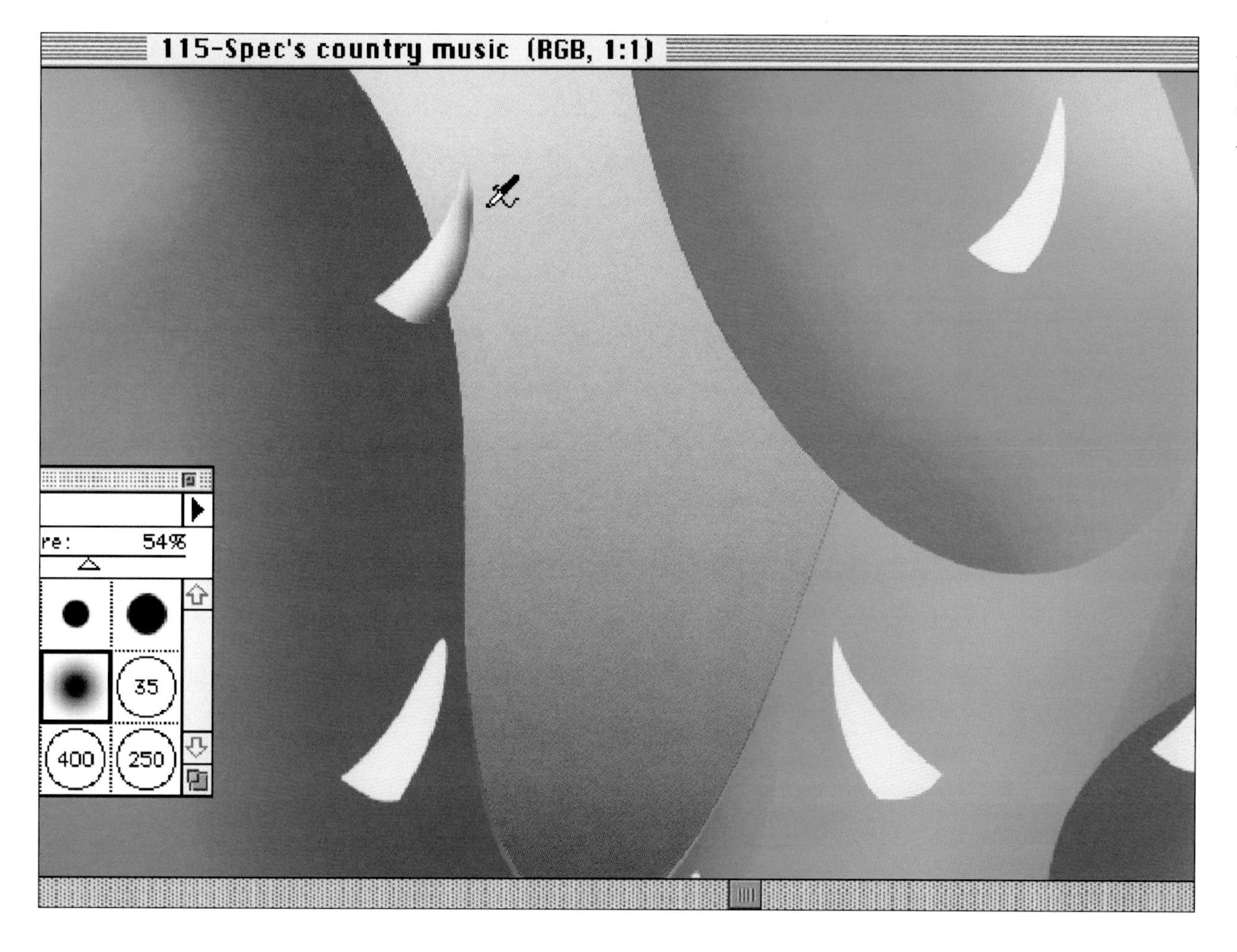

He adds final touches to the needles using the Airbrush tool.

Working in Photoshop: An Artist's Perspective

Before Adam invested in his computer system, he spoke with a lot of people about whether he should work in Photoshop or Adobe Illustrator. After much consideration, Adam finally decided that Photoshop was a little more "painterly" and intuitive.

He likes the fact that he can easily get so many different variations of color and texture, and that he can tear apart a composition, then rebuild it.

When I decided to go into commercial art, I learned airbrush for a year and this is actually a manual airbrush. And I'm really glad I did, because most of the way Photoshop works—and a lot of these programs—are all masking and they're all based on airbrush techniques a lot, more than, say, oil painting and brush work. Although you can use it that way too.

A lot of times when I have something, I sit here and I think

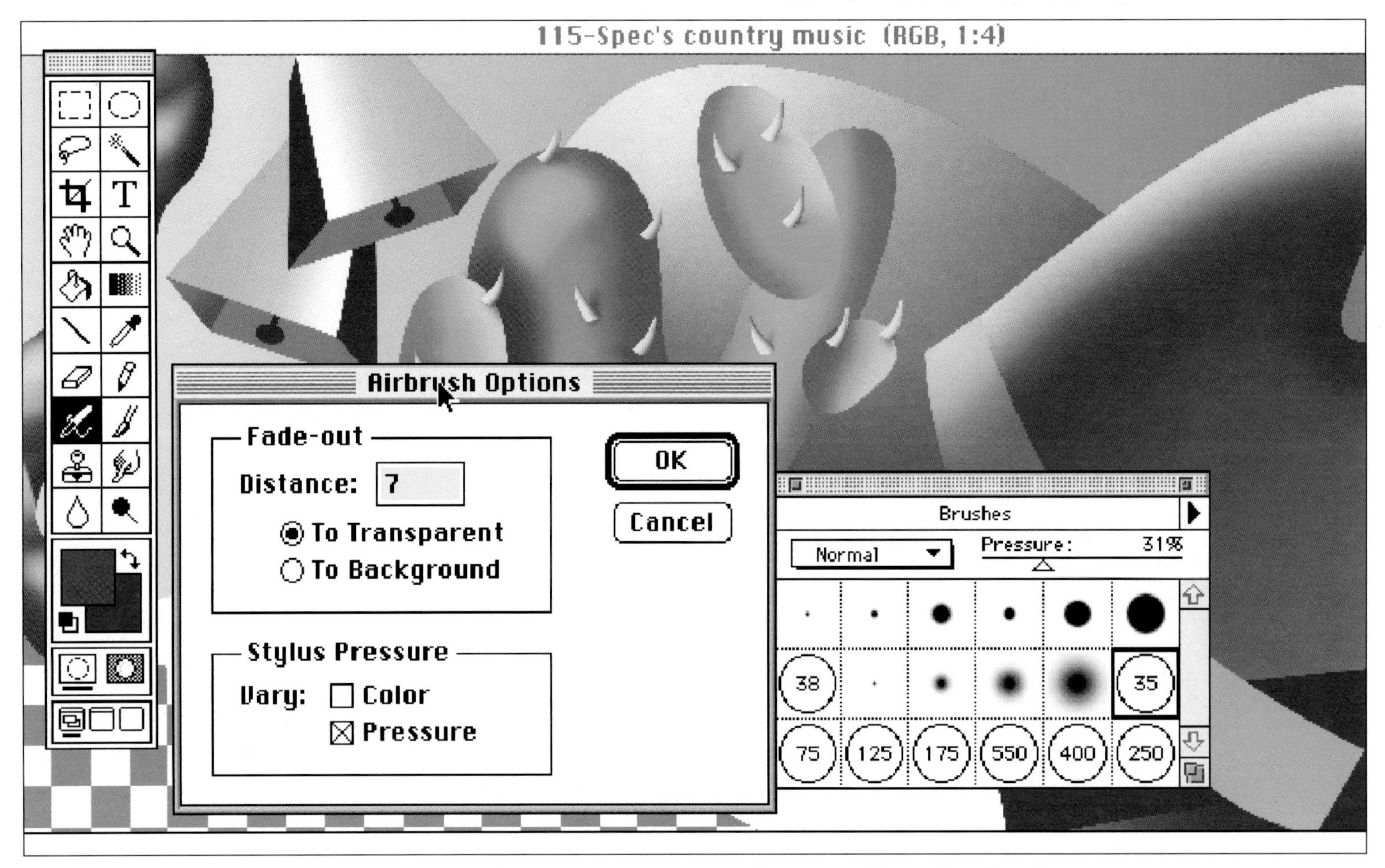

Using the Airbrush Options dialog box, Adam adjusts the transparency and pressure of the Airbrush tool.

"I'm not really sure how I'm going to do this". So I start like that every time. And then I discover so many things that I really should write notes down because I could use them for tips. But I take it for granted at this point that I just create what I need out of necessity. And the great thing is that you can do that with Photoshop. It's just like with an airbrush. You have to create a certain effect and you sit and think, "How can I make this happen?"

Adam doesn't always go for the quickest electronic option, such as cloning or mirroring. Sometimes he selects a warmer, more hand-drawn look, depending on how he wants a piece to look.

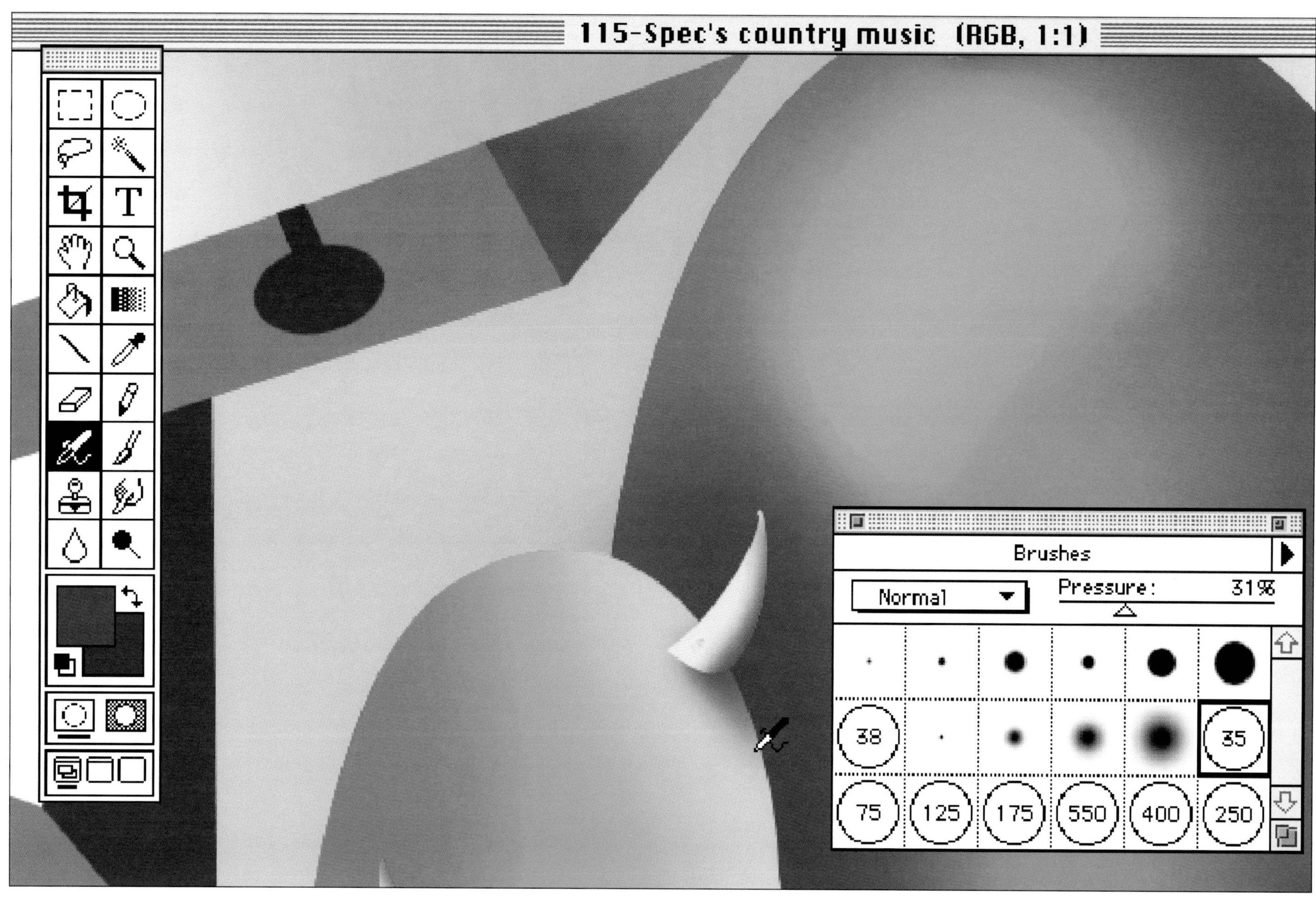

After creating all of the elements of the cactus, Adam adds some final touches. Here he sets the Paintbrush to fade out and adds shadows under the cactus needles.

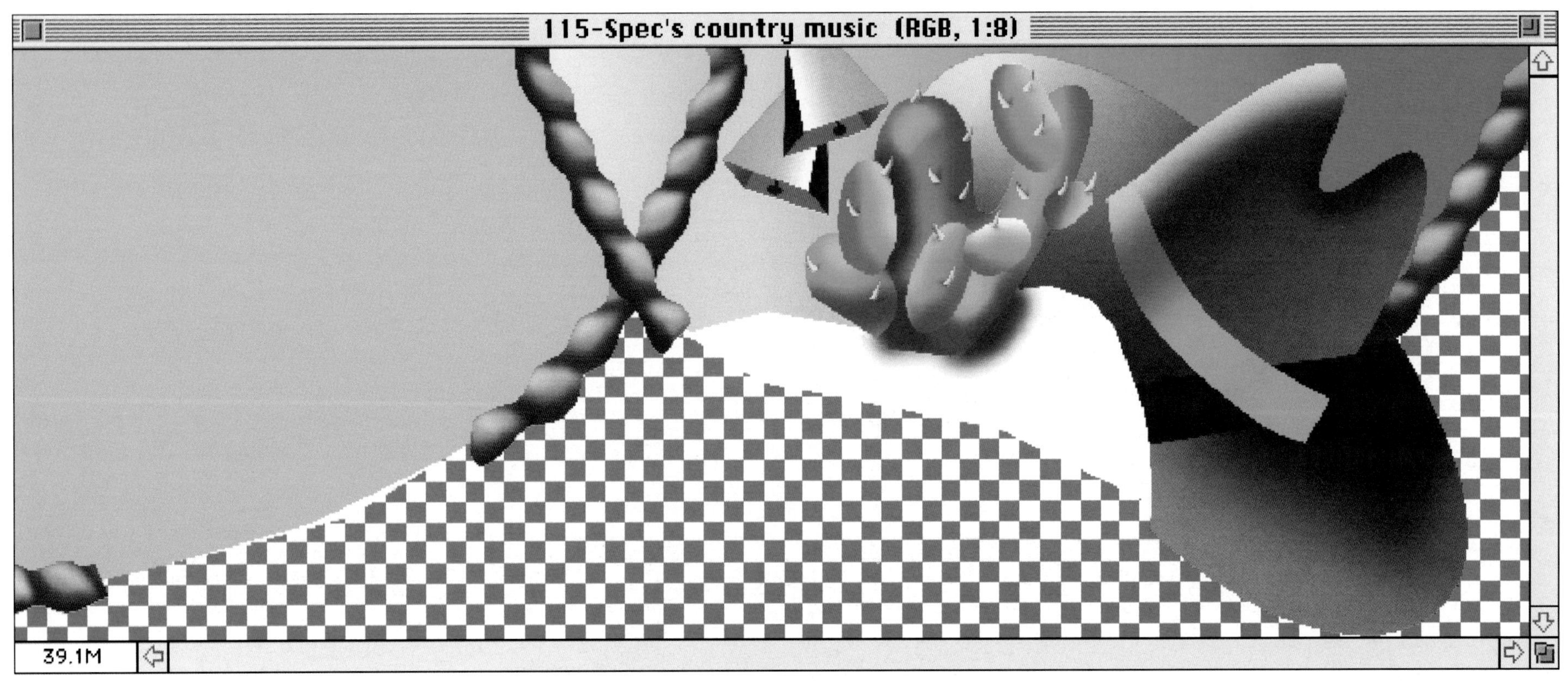

With the cactus completed, Adam can now begin adding the remaining elements—the guitar, the banjo, and the cowboy boots.

Regardless of how an element is created, Adam always skews things a little bit in his work. It creates a certain balance for him.

Another Photoshop element useful to Adam is the opacity he can get in his images. The painterly quality of Photoshop tools allows an artist to change the opacity of something so that it's not 100% color being placed. For example, Adam can highlight an object by adjusting the opacity of his color. Or, he can achieve the same effect with the Airbrush tool and a mask.

Adam combines the results he can get in Photoshop with the three-dimensional effects of RayDreams Designer. For example, to create the 3-D effect on the bat in one of his illustrations (see "Gallery," page 206), he used the simulated wood effect built into Ray Dreams Designer and changed the veining, vein direction, color, and highlight. He then created the shape of the bat, and imported it into Photoshop.

Adam created nine murals for Specs Music. The top image is for the Latin music section, the bottom for the video section.

Photo: Guy Salvador

EQUIPMENT

- Macintosh IIfx with a 175Mb internal hard drive, a Newer Technology accelerator, 128Mb RAM
- 19″ RasterOps Monitor with an accelerator card (2.88) and a 24-bit graphics board
- 44Mb SyQuest removable cartridge drive
- 500Mb Fujitsu external hard drive
- Microtek 300zs color scanner
- Wacom digitized pen and tablet (12″)
- Ricoh PC Laser Printer 6000
- Adobe Photoshop 2.5
- QuarkXPress 2.5
- RayDreams Designer
- Aldus Fetch
- Macromedia Director
- Andromeda Filters and Gallery Effects

CONVERSATIONS

Photo: Guy Salvador

ADAM COHEN
Illustrator

Adam Cohen, Illustrator, has a BFA in Fine Arts, and has had his traditional beachscapes and landscapes exhibited in many art galleries. His resumé includes electronic illustrations for a wide range of clients, including American Express, Estee Lauder, The Gap, Ziff Davis Publishing,* Ad Week, *Sony Music and Entertainment,* The Atlantic Monthly, Sports Illustrated for Kids, The New York Times, *and Henry Holt and Company.

I was at a party in Soho maybe eight or nine years ago and I remember walking in one room in the house that had a computer on with a big pen tablet. They left it on for people to play with this paint program. I was completed fascinated with it. I remember running up and trying to airbrush and asking "Why is the airbrush so grainy?" I think the programs at that time were not as sophisticated as they were when I finally decided to get into it. This must have been 1985. I remember being really fascinated with it.

I was using the airbrush and decided to go into commercial art and then I took a class at SVA [School of Visual Arts, New York] just to brush up on and get some concepts going with graphic designer Javier Romero. At the very end of the class he started saying, "The road to the future is computers."

I really felt that he was right. I started realizing that a lot of people weren't in on it. That was about four years ago and I spent a whole year researching before I bought a single piece of equipment. I found out everything, and spoke to so many people, and made sure that my style would interpret to computer before I did that. And then when I bought it, all of a sudden all these pieces arrived and came together, and I didn't have a clue of how it operated, how to work it. I had a tutor come and I watched some videotapes on how the Finder worked and windows worked. When that concept finally sunk in after a week, I think I did panic, thinking, "Oh my God, What did I do—I spent so much money. Is it really going to work for me?"

Then I had a client call to do a job and I had cold feet and I just did it. I did my first job for *Cosmopolitan* magazine on the computer doing a little spot for them about three years ago, in 1989. I was doing a whole series called "His Point of View." So from there on I just did it. I think the second one I did was the Corporate Scoreboard for *Business Week*. I remember talking with that art director about computers. I was talking to a lot of art directors and they were all telling me they would rather have a disk handed to them. They could just throw it into QuarkXPress. A lot of these art directors also were extremely new to it, and then some of them were just getting into it. It seemed like a lot of people in the industry were just starting to get into that in magazines and things. It was really that time when it was making that transition.

There were still a lot of illustrators who weren't on the computer. Now it seems there are so many more. It's like a huge magnet attracting so many people now.

Photo: Guy Salvador

SUSSAN GIALLOMBARDO
Creative Assistant

Sussan is a graduate of the Parson's School of Design, New York. She began painting and drawing when she was six, and hasn't stopped yet.

When I started working with Adam, I didn't know the computer at all. Now I've been working with it for about a year and a half. I was just helping him with design decisions and colors and making it faster for him to decide on something or adding an object or taking it away. Then I learned the computer and I can execute a piece and have more input on it, like if it should be feathered 30 or 40. It's like two scientists trying to work on something. It can get done a lot faster this way—instead of one person struggling for, say, eight hours, we can do it in three, and come up with something better. For this project, it would have been horrible to have one person working instead of two.

SUMMARY

About five years ago, people in the commercial art and design fields were talking about how computers are going to be the future of the industry. Adam Cohen took all of that talk seriously. He took a good look at his traditional fine arts background, and formulated a working plan for himself. He examined his assets and the new technology, then determined what needed to be done to bring him success. Then he jumped in. But only time will tell the total impact of computers on the art and design world.

GALLERY

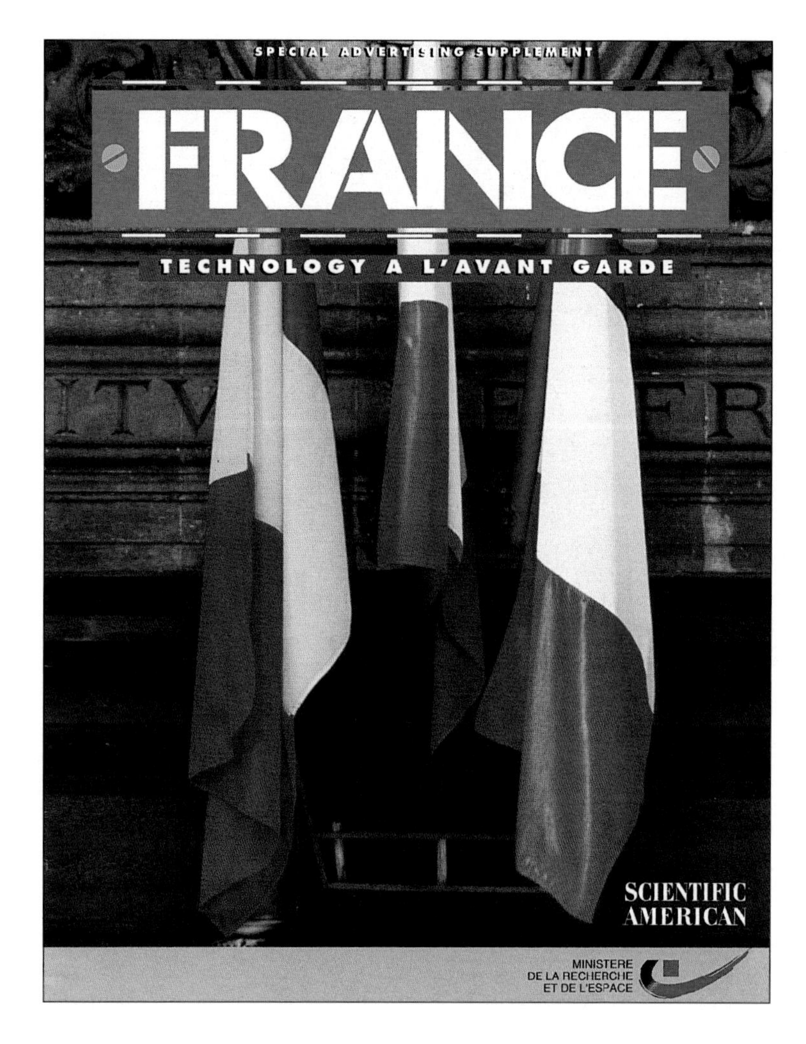

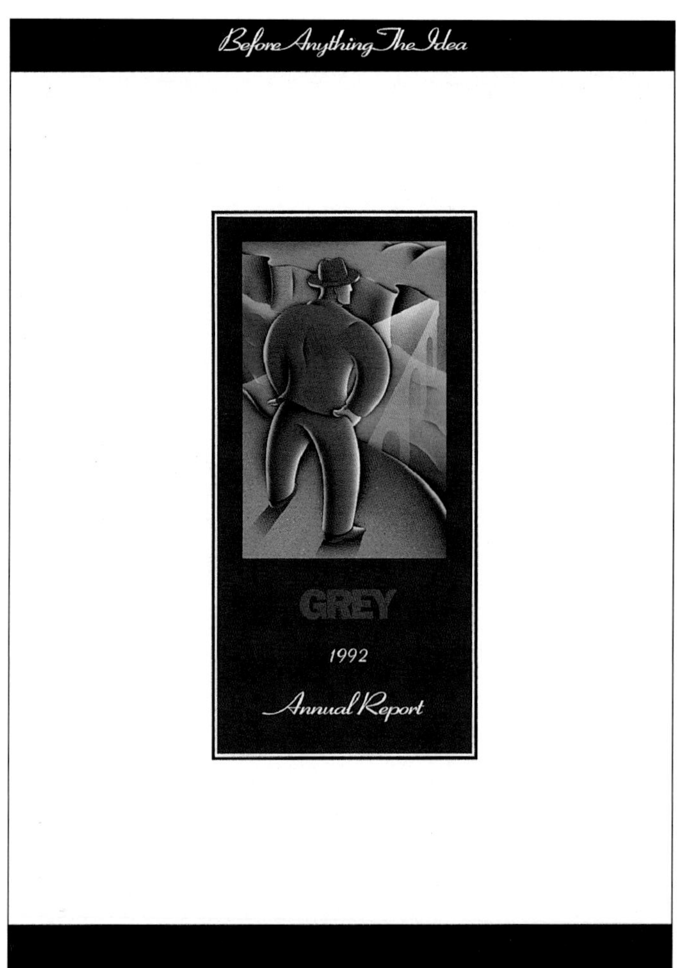

Left: Insert in Scientific American Magazine Right: 1993 annual report for Grey Advertising.

Left: Print ad for Food and Wine Magazine. Right: Multimedia brochure promoting the Multimedia group of shows

Left: Identity program for Hanbury Manor, a 17th century manor house in England, operated by Rockresorts.
Right: Logo, newsletter, signage, and menus for the Santa Fe Grill, a Hilton Hotels on-site restaurant.

Left: Identity program for The Lodge at Koele, Hawaii, operated by Rockresorts. Right: Diana DeLucia Design direct mail promotion.

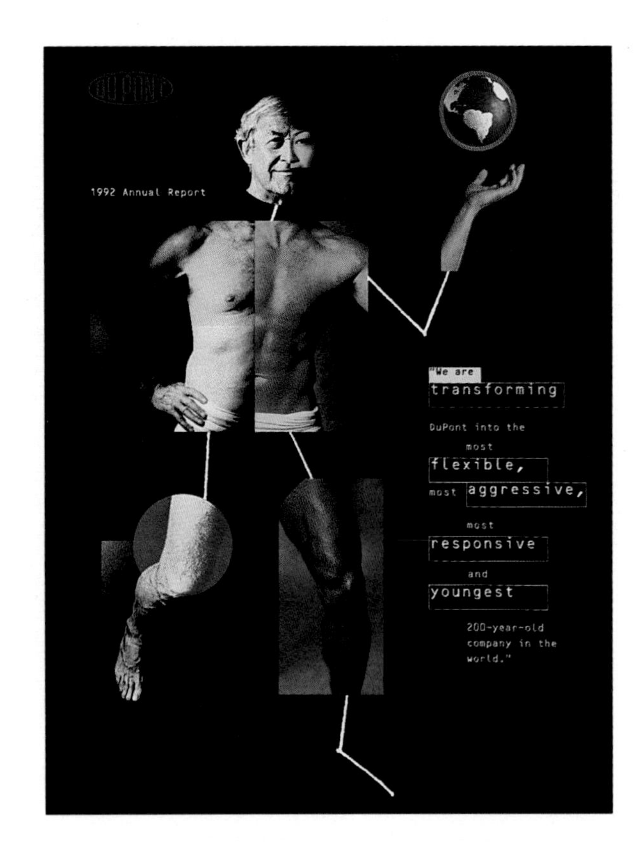

Left: 1992 annual report for DuPont. Right: Capabilities brochure for Toppan Printing Company.

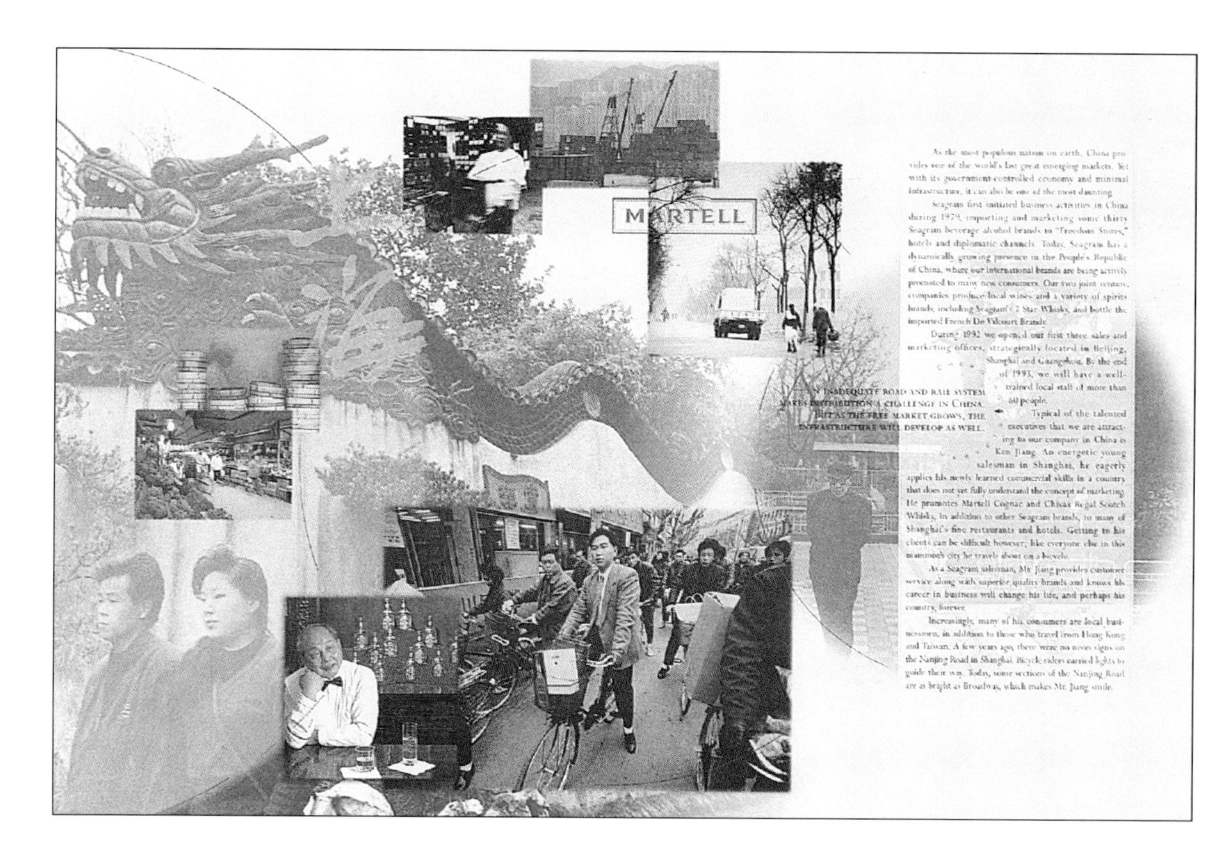

As the most populous nation on earth, China provides one of the world's last great emerging markets. Yet with its government-controlled economy and minimal infrastructure, it can also be one of the most daunting.

Seagram first initiated business activities in China during 1979, importing and marketing some thirty Seagram beverage alcohol brands to "Freedom Stores," hotels and diplomatic channels. Today, Seagram has a dynamically growing presence in the People's Republic of China, where our international brands are being actively promoted to many new consumers. Our two joint venture companies produce local wines and a variety of spirits brands, including Seagram's 7 Star Whisky, and bottle the imported French De Vilcourt Brandy.

During 1992 we opened our first three sales and marketing offices, strategically located in Beijing, Shanghai and Guangzhou. By the end of 1993, we will have a well-trained local staff of more than 60 people.

AN INADEQUATE ROAD AND RAIL SYSTEM MAKES DISTRIBUTION A CHALLENGE IN CHINA. BUT AS THE FREE MARKET GROWS, THE INFRASTRUCTURE WILL DEVELOP AS WELL.

Typical of the talented executives that we are attracting to our company in China is Ken Jiang. An energetic young salesman in Shanghai, he eagerly applies his newly learned commercial skills in a country that does not yet fully understand the concept of marketing. He promotes Martell Cognac and Chivas Regal Scotch Whisky, in addition to other Seagram brands, to many of Shanghai's fine restaurants and hotels. Getting to his clients can be difficult however; like everyone else in this mammoth city he travels about on a bicycle.

As a Seagram salesman, Mr. Jiang provides customer service along with superior quality brands and knows his career in business will change his life, and perhaps his country, forever.

Increasingly, many of his consumers are local businessmen, in addition to those who travel from Hong Kong and Taiwan. A few years ago, there were no neon signs on the Nanjing Road in Shanghai. Bicycle riders carried lights to guide their way. Today, some sections of the Nanjing Road are as bright as Broadway, which makes Mr. Jiang smile.

Left: 1992 annual report for The Seagram Company Ltd. Right: MTV media kit.

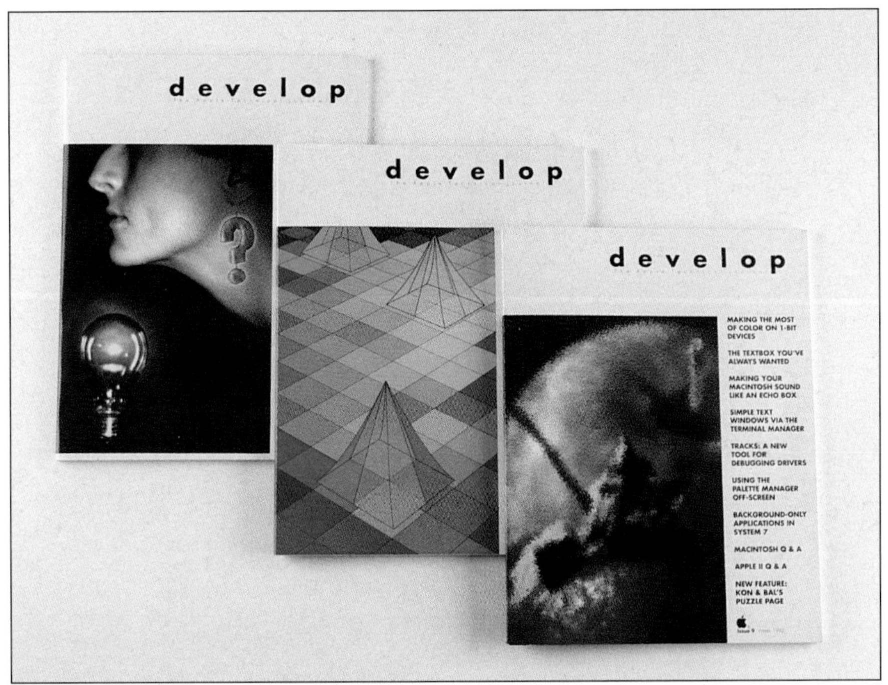

Left: Sleeves for RightBrain software. Right: Magazine cover for Apple Computer.

Left: Packaging for RayDreams addDepth software. Right: CD-ROM covers for Apple Computer.

Left: National and international magazine advertisement for Smirnoff through McCann Erickson. Right: Editorial art for "W" magazine (Fairchild Publications).

Left: National advetisement for Nobilia Watch through Young & Rubicam. Right: Assignment for Philips Lighting through McKinney & Silver.

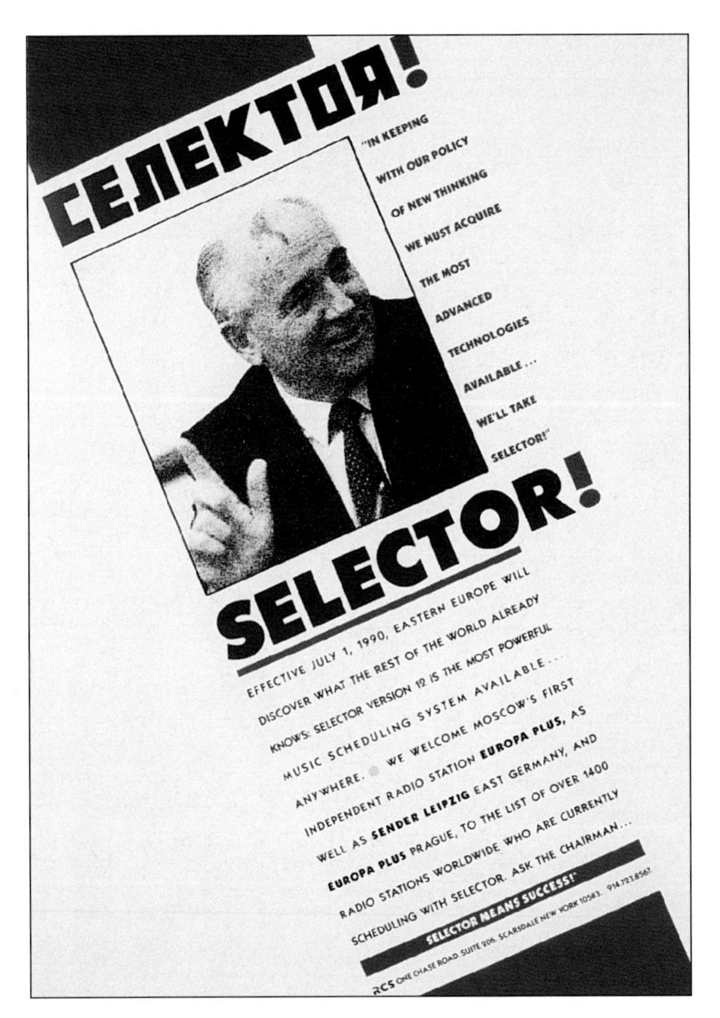

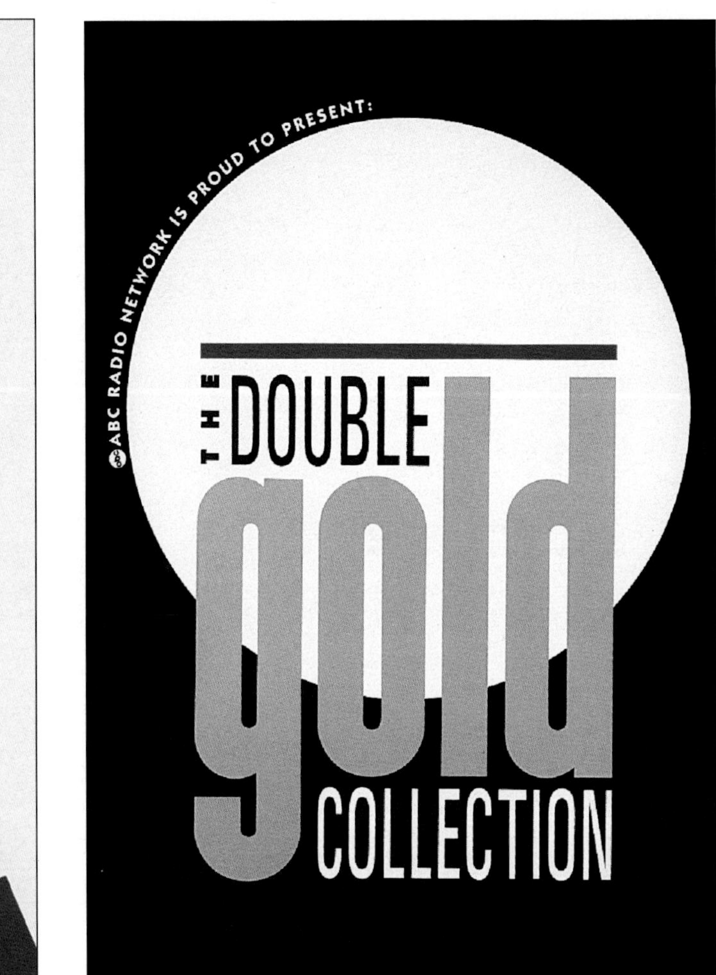

Left: Trade advertisement for Radio Computing Services. Right: Media kit cover for ABC Radio Network.

ONE *client station is all it takes to be in the music scheduling business. And, at RCS, even though we now have over a thousand client stations world wide, we never forget that when you're in a bind, you require the kind of attention you would get if you were our only client. We understand the importance of service. In fact, we built our business on it. When we say we won't leave you hanging... we mean it!* ***TEN*** *years in the radio industry is a very long time. SELECTOR has the distinction of having been a stable force in programming since its release in 1979. That's a full decade of innovation and improvement packed into a music scheduling system that has been on the front lines of music programming right from the beginning. We've dedicated ourselves to offering a system that reflects the needs of our clients. And, we're looking forward to our second decade of growth, innovation and service.* ***ONE THOUSAND*** *radio stations world-wide are now using SELECTOR, making it the standard by which all other music systems are judged. When the time comes for you to evaluate your music scheduling system, consider the fact that more radio stations have chosen SELECTOR than any other system, anywhere at any price. Call us for a free SELECTOR demo disk and see for yourself what over 1000 radio stations around the world already know... SELECTOR means success!*

RCS RADIO COMPUTING SERVICES, INC.

One Chase Road Suite 206 Scarsdale, New York 10583 (513) 941 0829 (914) 723 8567 FAX (914) 723 6651

Left: Trade advertisement for Radio Computing Services. Right: Business-to-business brochure for Radio Computing Services.

Left: Editorial art for "We Deliver," a takeout menu magazine. Right: Editorial art for "Rookie League," a children's baseball magazine (Welsh Publishing Group)..

Left: Illustration for Inquiry & Investigation: Cities *(Word and Image Design). Right: Book cover art for* Drug Store Cowboy *(Dell Books).*

GLOSSARY

airbrushing

A hand-held ink sprayer that can create different ink textures. The textures can be smooth, gradual, even, or light to dark.

alpha channel

A 32-bit color system used to mask areas for transparencies, overlays, and special effects.

analog system

Analog systems monitor conditions, such as sound and movement, then convert them to analogous electronic or mechanical patterns. For example, an analog watch converts the movement of the Earth into rotations of hands. Analog systems are contrasted with digital systems, which break conditions into numbers.

bitmap

A graphic image that is represented by individual dots (pixels) laid out on a grid.

camera-ready copy

Graphics, illustration, text, etc. in its final form ready to be photographed for reproduction.

CD-ROM

(*compact disc read-only memory*) The CD-ROM holds 600Mb of text, graphics and/or stereo sound (similar to a music CD).

chromalin

Translucent color photographic film, sometimes called a transparency or a chrome.

CMYK color

A color model using cyan (C), magenta (M), yellow (Y), and black (K) as the basic inks to form different colors. Generally used for color separations.

color key

A method, developed by the 3M Company, used to show progressive color breakdown. The resulting proofs are useful for checking registration, size, and blemishes. They are not a good indication for checking actual color.

comp

Abbreviation for composition layout. A mock-up of a design to be used by the client or designer.

continuous tone

An unscreened photograph or illustration containing gradient tones from black to white.

cyan

A greenish blue color used in process color printing.

desktop publishing

Using computers to produce high quality text and graphics output to be sent to commercial printers. The common abbreviation is *DTP*.

digital systems

Accept and process data that has been converted into numbers. For example, digital systems represent data as binary numbers-series of ones and zeros.

dot gain

A defect in printing when temperature, ink, and paper type cause an increase in each drop of ink.

double-burning

A process by which two images are imposed on each other for the purpose of creating one image.

DPI

(*dots per inch*) A printer resolution measurement.

dye sublimation print

A color printing method that uses dye, instead of ink, to produce continuous color that approaches photographic quality.

dye transfer

A continuous tone print produced from a transparency.

EPS

(*encapsulated PostScript*) A high-resolution graphic format that allows you to manipulate and preview your image on screen.

feathering

Blending the edges of an object in an irregular way. This blending can be done into another object or the background to achieve the desired effect.

FPO

The placing of art work "for position only" and not for reproducing.

halftone

A continuous image made by a screen that causes the image to be broken into various size dots. Smaller dots produce lighter areas and larger dots produce darker areas.

hot metal type

Type that has been set by machine or hand in cast metal.

ink-jet printing

As paper moves through the printer rows of minute jets squirt ink to form an image.

kerning

Adjusting the white space between two characters to make a word appear and read better.

magenta

A bluish red color used in process color printing.

masking

Blocking out part of an image to deselect it, to get rid of it, to get rid of unwanted details, or add to it.

mechanical

Camera-ready art that has been physically put together by cut-and-paste methods.

multimedia

Information from several sources using graphics, text, audio, and full-motion video.

opacity

The amount of ink that shows through onto the other side of the paper.

paste-up

The activity of physically preparing the camera-ready art.

pen tablet

A graphics drawing tablet used for sketching, drawing, and painting in conjunction with drawing or painting software.

pixel

Short for picture element. The smallest element or dot that can be seen on a computer screen.

plates

The actual thin sheet of metal or plastic on a printing press containing the image to be printed. When it is inked, it produces the printed master.

PMS

(*Pantone Matching System*) A worldwide system of standardized ink colors used to specify and check color.

PostScript

A computer language used to describe images and type for laser printers and other output devices developed and trademarked by Adobe Systems, Inc.

prepress

The complete preparation of camera-ready materials up to printing.

process color

Also called four-color process. Mixing the four standard printing color inks (CMYK-cyan [C], yellow [Y], magenta [M], and black [K]) to create images. A transparency is made for each individual color and the color effect is done by overlapping the four transparent ink colors.

RAM

(*random access memory*) The amount of space your computer provides to temporarily store information. RAM space is considered volatile, as anything stored in it is erased when your computer is turned off.

resample

Changing an image's resolution.

resolution

Resolution is the number of dots or pixels per inch used by an output device.

RGB color

A method of simulating colors by combining the colors red (R), green (G) and blue (B).

RIP

(*raster image processor*) Prepares data for an output device, usually a printer.

ROM

(*read only memory*) Permanently stores instructions and programs that cannot be altered.

scanning

Converting line art, photographs, text, or graphic from paper to a bit-mapped image for manipulation and placement that can be used on a computer.

spot color

Color that is applied only to a specific area. At the time of color separation the spot color is assigned its own color plate.

stripping

The physical placement of photographs, illustrations, text, graphics, and color areas in preparation to making a plate.

stylus

A pressure-sensitive pen-like instrument that enables you to control the rate of flow from a tool in a particular drawing program.

TARGA file

A file format that is most common in higher-end PC-based paint systems.

template

A predesigned pattern used as a format to create new documents.

thermal wax printing

A medium-resolution printing process that transfers wax-like ink onto paper.

TIFF

(*tagged image file format*) A versatile graphics format, developed by Aldus Corp. and Microsoft Corp., which stores a gray map that specifies the location and level of gray associated with each pixel.

tracking

The adjustment of space between more than two characters in a given section of text.

transparency

Color translucent photographic film, sometimes called a *chrome* or *chromalin*.

trapping

An overlapping technique that allows for misregistration of the color plate to prevent gaps in color.

typography

The appearance and control of characters that make up type.

veloxes

A halftone screened back-and-white print. Also the tradename for a printing paper made by Kodak.

wire system

An electronic system over which news and information is sent to television stations, radio stations, newspapers, and other newsrooms.

INDEX

D